METHODS IN MOLECULAR BIOLOGY

Series Editor
John M. Walker
School of Life and Medical Sciences
University of Hertfordshire
Hatfield, Hertfordshire, AL10 9AB, UK

For further volumes:
http://www.springer.com/series/7651

Thermogenic Fat

Methods and Protocols

Edited by

Jun Wu

Department of Molecular and Integrative Physiology, Life Sciences Institute, University of Michigan, Ann Arbor, MI, USA

Humana Press

Editor
Jun Wu
Department of Molecular and Integrative
Physiology, Life Sciences Institute
University of Michigan
Ann Arbor, MI, USA

ISSN 1064-3745 ISSN 1940-6029 (electronic)
Methods in Molecular Biology
ISBN 978-1-4939-6819-0 ISBN 978-1-4939-6820-6 (eBook)
DOI 10.1007/978-1-4939-6820-6

Library of Congress Control Number: 2017932076

Printed on acid-free paper

This Humana Press imprint is published by Springer Nature
The registered company is Springer Science+Business Media LLC
The registered company address is: 233 Spring Street, New York, NY 10013, U.S.A.

Preface

The rediscovery of thermogenic adipocytes in human adults has reignited a great deal of enthusiasm towards these unique metabolic cells that convert chemical energy into heat, therefore holding great promise for helping to counteract obesity and its associated metabolic disorders. It is now appreciated that at least two types of thermogenic adipocytes exist in rodents and humans, classical brown and inducible beige adipocytes. These two types of fat cells arise from distinct developmental origins and position in different anatomical locations. In contrast to white adipocytes that store surplus energy, brown and beige fat cells dissipate chemical energy through uncoupling protein 1 (UCP1), together with other mechanisms. In addition to their respective roles in mediating energy balance, all adipocytes function as endocrine organs and directly control systemic metabolism through adipocyte-secreted hormones (so-called adipokines) and metabolites. It is now considered settled science that adipose tissue as a whole plays a key role in maintaining metabolic homeostasis. Emerging evidence suggests that the mass and activity of thermogenic fat influence body weight, insulin sensitivity, glucose tolerance, and other key metabolic parameters. Further investigation and deeper understanding of the development and function of these thermogenic fat cells may lead to novel therapeutic strategies against metabolic syndromes.

Targeting both experts in this area who want to compare notes with colleagues in the field and newcomers that look forward to learning assays and starting their own investigation of brown and beige fat function, this volume collects protocols that describe methodologies to study thermogenic fat biology from various angles. The first part focuses on how to establish in vitro culture systems. Historically, many adipocyte-related studies have been carried out using immortalized cell lines, for example the famous 3T3-L1 cell line that was originally established by Dr. Howard Green's group. While cell line work remains to be an indispensible approach, much of the thermogenic fat specific function can only be evaluated in primary cultures. Several chapters introduced methods on how to isolate, culture, and differentiate primary fat cells from both laboratory mice and humans. It has been reported that immune cells within the subcutaneous fat tissue regulate thermogenic fat functions through paracrine mechanisms. Several chapters presented flow cytometry methods to isolate various subpopulations of precursors within the stromal vascular fraction (SVF) of the adipose tissue, which contains both preadipocytes and immune cells. In the second part of the book, we first introduced multiple means to genetically manipulate and evaluate brown and beige fat in vivo. Since one of the most prominent functions of thermogenic fat is to mediate energy consumption through mitochondrial respiration, two chapters were dedicated to discuss methods on bioenergetics analyses both in vitro and in vivo. We further presented how to experimentally study ion channel function and cell-cell communication in these cells. Lastly, we discussed how to evaluate thermogenic fat content and activity in humans, how to culture and assay these cells through interdisciplinary approaches, including 3-D adipospheres and microfluidic systems, and how to use thermogenic fat cell lines to carry out drug screens.

The editor would like to thank all the contributing authors who shared their "trade secrets" here without reservations so that readers did not have to "reinvent the wheel" and learn these key assays with much less challenge. I am also very grateful to Dr. John Walker, who commissioned this timely book and has provided much invaluable advice along the way.

Ann Arbor, MI, USA ***Jun Wu***

Contents

Contributors

TAKESHI AKAMA • *Department of Internal Medicine, Division of Metabolism, Endocrinology & Diabetes, University of Michigan Medical School, Ann Arbor, MI, USA; Biointerfaces Institute, University of Michigan, Ann Arbor, MI, USA*
ROHOLLAH BABAEI • *DKFZ Junior Group Metabolism and Stem Cell Plasticity, German Cancer Research Center, Heidelberg, Germany*
NICKI A. BAKER • *Section of General Surgery, Department of Surgery, University of Michigan Medical School, Ann Arbor, MI, USA*
IREM BAYINDIR-BUCHHALTER • *DKFZ Junior Group Metabolism and Stem Cell Plasticity, German Cancer Research Center, Heidelberg, Germany*
JONATHAN R. BRESTOFF • *Department of Pathology and Immunology, Washington University School of Medicine, Saint Louis, MO, USA*
JESSICA C. BROOKS • *Department of Chemistry and Biochemistry, Auburn University, Auburn, AL, USA*
ZHIYONG CHENG • *Department of Human Nutrition, Foods and Exercise, Fralin Life Science Institute, College of Agriculture and Life Science, Virginia Tech, Blacksburg, VA, USA*
MARIA CHONDRONIKOLA • *Center for Human Nutrition, School of Medicine, Washington University in St. Louis, St. Louis, MO, USA; Harokopio University of Athens, Athens, Greece*
TAE-HWA CHUN • *Department of Internal Medicine, Division of Metabolism, Endocrinology & Diabetes, University of Michigan Medical School, Ann Arbor, MI, USA; Biointerfaces Institute, University of Michigan, Ann Arbor, MI, USA*
PAUL COHEN • *Laboratory of Molecular Metabolism, The Rockefeller University, New York, NY, USA*
SARAH R. DONNELLY • *Department of Human Nutrition, Foods and Exercise, Fralin Life Science Institute, College of Agriculture and Life Science, Virginia Tech, Blacksburg, VA, USA*
CHRISTOPHER J. EASLEY • *Department of Chemistry and Biochemistry, Auburn University, Auburn, AL, USA*
MARGO P. EMONT • *Department of Molecular and Integrative Physiology, Life Sciences Institute, University of Michigan, Ann Arbor, MI, USA*
WEI GAO • *Department of Geriatrics, The Second Affiliated Hospital of Nanjing Medical University, Nanjing, China; Department of Medicine, Physiology and Biophysics, Center for Diabetes Research and Treatment, Center for Epigenetics and Metabolism, University of California-Irvine, Irvine, CA, USA*
NICOLÁS GÓMEZ-BANOY • *Metabolic Health Center, Division of Cardiology, Department of Medicine, Weill Cornell Medicine, New York, NY, USA*
ANTONIA GRAJA • *Research Group Adipocyte Development, German Institute of Human Nutrition (DIfE), Nuthetal, Germany*
ANNE-MARIE JANK • *Research Group Adipocyte Development, German Institute of Human Nutrition (DIfE), Nuthetal, Germany*

ROBERT L. JUDD • *Department of Anatomy Physiology and Pharmacology, Auburn University, Auburn, AL, USA*
LAWRENCE KAZAK • *Dana-Farber Cancer Institute, Harvard Medical School, Cancer Biology CLS 11116, Boston, MA, USA*
EUN RAN KIM • *Brown Foundation Institute of Molecular Medicine, University of Texas McGovern Medical School, Houston, TX, USA*
XINGXING KONG • *Division of Endocrinology, Diabetes and Metabolism, Beth Israel Deaconess Medical Center, Harvard Medical School, Harvard University, Boston, MA, USA*
JOE LABUZ • *Department of Biomedical Engineering, University of Michigan, Ann Arbor, MI, USA; Biointerfaces Institute, University of Michigan, Ann Arbor, MI, USA*
YUN-HEE LEE • *College of Pharmacy, Yonsei University, Incheon, South Korea*
BRENDAN M. LEUNG • *Department of Biomedical Engineering, University of Michigan, Ann Arbor, MI, USA; Biointerfaces Institute, University of Michigan, Ann Arbor, MI, USA*
LONGHUA LIU • *Department of Human Nutrition, Foods and Exercise, Fralin Life Science Institute, College of Agriculture and Life Science, Virginia Tech, Blacksburg, VA, USA*
TIEMIN LIU • *Division of Hypothalamic Research, The University of Texas Southwestern Medical Center at Dallas, Dallas, TX, USA*
JAMES C. LO • *Metabolic Health Center, Division of Cardiology, Department of Medicine, Weill Cornell Medicine, New York, NY, USA*
CAREY N. LUMENG • *Department of Pediatrics and Communicable Diseases, University of Michigan Medical School, Ann Arbor, MI, USA*
IRINA MELN • *DKFZ Junior Group Metabolism and Stem Cell Plasticity, German Cancer Research Center, Heidelberg, Germany*
ANIL MISHRA • *Division of Cardiovascular Medicine, Department of Internal Medicine, Carver College of Medicine, University of Iowa, Iowa City, IA, USA*
LINDSEY A. MUIR • *Department of Pediatrics and Communicable Diseases, University of Michigan Medical School, Ann Arbor, MI, USA*
SEAN O'CONNOR • *Laboratory of Molecular Metabolism, The Rockefeller University, New York, NY, USA*
JOHN O. OGUNBILEJE • *Department of Surgery, University of Texas Medical Branch at Galveston, Galveston, TX, USA; Metabolism Unit, Shriners Hospitals for Children-Galveston, Galveston, TX, USA*
ROBERT W. O'ROURKE • *Section of General Surgery, Department of Surgery, University of Michigan Medical School, Ann Arbor, MI, USA; Department of Surgery, Ann Arbor Veteran's Administration Hospital, Ann Arbor, MI, USA*
CRAIG PORTER • *Department of Surgery, University of Texas Medical Branch at Galveston, Galveston, TX, USA; Metabolism Unit, Shriners Hospitals for Children-Galveston, Galveston, TX, USA*
SUSHIL G. RANE • *Cell Growth and Metabolism Section, Diabetes, Endocrinology, and Obesity Branch, NIDDK, National Institutes of Health, Clinical Research Center, Bethesda, MD, USA*
RAJAN SAH • *Division of Cardiovascular Medicine, Department of Internal Medicine, Carver College of Medicine, University of Iowa, Iowa City, IA, USA; Fraternal Order of the Eagles Diabetes Research Center, Iowa City, IA, USA*
ISAAC SAMUEL • *Department of Surgery, University of Iowa, Carver College of Medicine, Iowa City, IA, USA*

TIM J. SCHULZ • *Research Group Adipocyte Development, German Institute of Human Nutrition (DIfE), Nuthetal, Germany; German Center for Diabetes Research (DZD), München-Neuherberg, Germany*
FARNAZ SHAMSI • *Section on Integrative Physiology and Metabolism, Joslin Diabetes Center, Harvard Medical School, Boston, MA, USA*
LABROS S. SIDOSSIS • *Harokopio University of Athens, Athens, Greece; Department of Surgery, University of Texas Medical Branch at Galveston, Galveston, TX, USA; Metabolism Unit, Shriners Hospitals for Children-Galveston, Galveston, TX, USA; Departments of Medicine, Rutgers University, New Brunswick, NJ, USA; Department of Kinesiology and Health, Rutgers University, New Brunswick, NJ, USA*
JESSICA K. SMITH • *Department of Surgery, University of Iowa, Carver College of Medicine, Iowa City, IA, USA*
JOCHEN STEINBRING • *Research Group Adipocyte Development, German Institute of Human Nutrition (DIfE), Nuthetal, Germany*
SHUICHI TAKAYAMA • *Department of Biomedical Engineering, University of Michigan, Ann Arbor, MI, USA; Biointerfaces Institute, University of Michigan, Ann Arbor, MI, USA*
DAN TONG • *Division of Cardiovascular Medicine, Department of Internal Medicine, Carver College of Medicine, University of Iowa, Iowa City, IA, USA*
QINGCHUN TONG • *Brown Foundation Institute of Molecular Medicine and Graduate Program in Neuroscience of Graduate School of Biological Sciences, University of Texas McGovern Medical School, Houston, TX, USA; Department of Neurobiology and Anatomy, University of Texas McGovern Medical School, Houston, TX, USA*
YU-HUA TSENG • *Harvard Stem Cell Institute, Harvard University, Cambridge, MA, USA*
ALEXANDROS VEGIOPOULOS • *DKFZ Junior Group Metabolism and Stem Cell Plasticity, German Cancer Research Center, Heidelberg, Germany*
UMESH D. WANKHADE • *Cell Growth and Metabolism Section, Diabetes, Endocrinology, and Obesity Branch, NIDDK, National Institutes of Health, Clinical Research Center, Bethesda, MD, USA*
KEVIN W. WILLIAMS • *Division of Hypothalamic Research, The University of Texas Southwestern Medical Center at Dallas, Dallas, TX, USA*
JUN WU • *Department of Molecular and Integrative Physiology, Life Sciences Institute, University of Michigan, Ann Arbor, MI, USA*
LING WU • *Department of Biochemistry, Center for Drug Discovery, and Translational Obesity Research Center, Virginia Tech, Blacksburg, VA, USA*
LITAO XIE • *Division of Cardiovascular Medicine, Department of Internal Medicine, Carver College of Medicine, University of Iowa, Iowa City, IA, USA*
BIN XU • *Department of Biochemistry, Center for Drug Discovery, and Translational Obesity Research Center, Virginia Tech, Blacksburg, VA, USA*
QIN YANG • *Department of Medicine, Physiology and Biophysics, Center for Diabetes Research and Treatment, Center for Epigenetics and Metabolism, University of California-Irvine, Irvine, CA, USA*
YANHUI ZHANG • *Division of Cardiovascular Medicine, Department of Internal Medicine, Carver College of Medicine, University of Iowa, Iowa City, IA, USA*
LOUISE D. ZHENG • *Department of Human Nutrition, Foods and Exercise, Fralin Life Science Institute, College of Agriculture and Life Science, Virginia Tech, Blacksburg, VA, USA*

Part I

Isolation and Culture Methods

Chapter 1

Isolation, Primary Culture, and Differentiation of Preadipocytes from Mouse Brown Adipose Tissue

Wei Gao, Xingxing Kong, and Qin Yang

Abstract

Evolutionally, brown adipose tissue (BAT) is developed for nonshivering thermogenesis to prevent hypothermia. BAT has a high capacity to dissipate chemical energy generated from metabolism of nutrients for heat production. Therefore when BAT is activated, nutrients are "burned' instead of being stored. This feature makes BAT an attractive target for obesity treatment. To investigate BAT function and regulation, brown adipocyte culturing is indispensable. This chapter describes a detailed protocol for isolation, primary culture, and differentiation of preadipocytes from mouse BAT. The preadipocytes can be used for investigating the regulation of brown fat cell differentiation. The differentiated brown adipocytes maintain major BAT features including high expression of uncoupling protein-1 and can be used to study BAT biology and pharmacology.

Key words Brown adipose tissue, Preadipocytes, Isolation, Culture, Differentiation

1 Introduction

Increasing energy expenditure to exceed energy intake and create a negative energy balance is a fundamental strategy for obesity treatment [1]. Brown adipose tissue (BAT) has been a primary focus for obesity research due to its significant capacity to dissipate food energy for heat production [2–4]. BAT expresses a high level of uncoupling protein-1 (UCP1), a mitochondrial transmembrane protein, which catalyzes a proton leak across the inner mitochondrial membrane [5, 6]. This leads to the reduction of proton gradient between mitochondrial intermembrane space and mitochondrial matrix. Since such a proton gradient is essential for oxidative phosphorylation in the mitochondria, UCP1 diverts fuel oxidation from ATP synthesis to heat production.

It has been long known that BAT plays a major role in nonshivering thermogenesis in rodents [7]. In humans, it was initially thought that BAT exists only in infants, likely also for nonshivering thermogenesis (NTS) and hypothermia prevention [8, 9]. While the infants grow up, NTS becomes nonessential and white adipose

Jun Wu (ed.), *Thermogenic Fat: Methods and Protocols*, Methods in Molecular Biology, vol. 1566,
DOI 10.1007/978-1-4939-6820-6_1, © Springer Science+Business Media LLC 2017

tissue gradually replaces BAT with age. However, mounting evidence from the research over the past several years clearly shows that the metabolically active BAT is presented in adult humans [10–13]. The findings triggered more enthusiasms for brown fat research. Potentially stimulating BAT activity could be an important strategy to increase energy expenditure for obesity treatment. In fact, several pharmacological approaches have been tested to stimulate BAT activity for obesity treatment [14, 15].

To study BAT biological function and pharmacological effects, in vitro brown adipocyte culture is indispensable [16–18]. Mature brown fat cells are terminally differentiated and there is no established method to culture them. BAT contains not only mature adipose cells, but also stromal vascular fraction (SVF), which includes preadipocytes [19–21]. SVF can be separated from mature adipocytes by collagenase digestion and preadipocytes can be further induced to differentiate to mature brown adipocytes for downstream applications such as genetic intervention or pharmacological stimulation. We here provide a detailed protocol for isolation, primary culture, and differentiation of preadipocytes from mouse BAT.

2 Materials

2.1 Animal

Newborn mice (1–2 day old) (*see* **Notes 1** and **2**).

2.2 Tools

Scissors and forceps (2 sets), 10 cm dishes, 50 mL Falcon tubes, 15 mL Falcon tubes, 100 μm cell strainers, 220 μm syringe driven filters, 50 mL syringes, cell culture plates (6–24 well), water bath.

2.3 Buffer and Culture Medium

1. Isolation buffer: 123.0 mM NaCl, 5.0 mM KCl, 1.3 mM $CaCl_2$, 5.0 mM Glucose, 100.0 mM N-2-hydroxyethylpiperizine-N'-2-ethanesulfonic acid (HEPES), 1×Penicillin-Streptomycin, 4 % Bovine serum albumin (BSA). Add ddH_2O to a volume of 100 mL. Mix and adjust pH with 7.45. Dispense into 15 mL tubes (10 mL/tube) and store at −20 °C.
2. Digestion buffer: Make 1 mL for every 2 mice. Thaw isolation buffer and add Collagenase II with a final concentration of 1 mg/mL (*see* **Note 3**). Then vortex and filter the buffer through 220 μm filters into 15 mL or 50 mL tube. Keep the buffer in 37 °C water bath.
3. Primary brown preadipocytes culture medium: Add 75 mL Fetal bovine serum (FBS) and 5 mL 100×Penicillin-Streptomycin into 420 mL Dulbecco's Modified Eagle Medium/Nutrient Mixture F-12 (DMEM/F12), GlutaMAX™ supplement.
4. Induction medium: Dissolve 0.5 mM 3-Isobutyl-1-methylxanthine (IBMX), 1 μM Dexamethasone, 1 μg/mL Insulin, 1 μM Rosiglitazone, 1 nM 3,3′,5-Triiodo-L-thyronine (T3) in primary brown preadipocytes culture medium. Filter through 220 μm filters.

5. Maintain medium: Dissolve 1 μg/mL insulin and 1 nM T3 in primary brown preadipocytes culture medium. Filter through 220 μm filters.

3 Methods

3.1 Isolation of Primary Brown Preadipocytes

1. Autoclave scissors and forceps (2 sets: one for dissection; the other for mincing BAT tissue) at least one day before preadipocyte isolation.
2. Immediately after decapitation, mouse is placed with the backup. Cut open along the back midline from the neck, and the two lobes of brown fat pad can be visible between the shoulders (interscapular region) (Fig. 1a). Carefully dissect out the brown adipose tissue, and remove the surrounding white fat and muscle (Fig. 1b). Place the tissue in a 10 cm culture dish on ice, with ice-cold Phosphate Buffered Saline (PBS) (*see* **Note 2**).
3. After finishing the dissection, aspirate PBS as much as possible. Then mince the tissue into small pieces (approximately 0.5–1 mm^3, or smaller) by using the second set of autoclaved scissors and forceps.
4. Transfer the minced tissue into a 50 mL tube containing digestion buffer (1 mL for every two mice), add equivalent volume of PBS, vortex for 10 s.

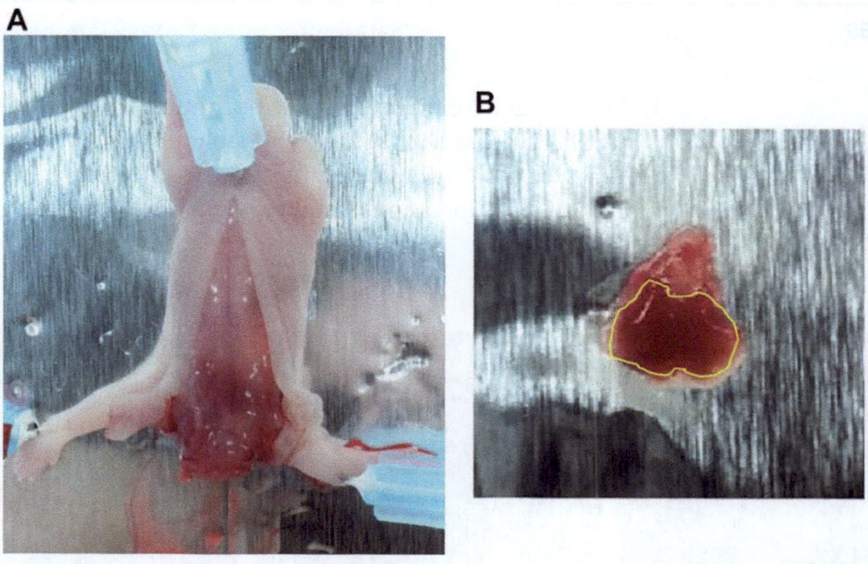

Fig. 1 Dissecting brown fat from a newborn mouse. (**a**) predissection, *brown fat* can be visualized in the interscapular region within the *yellow circle*. (**b**) dissected brown fat pad. The brown fat needs to be further cleaned by removing the surrounding muscle and white adipose tissue

5. Digestion: Place the tube into 37 °C shaking water bath at 150 cycles/minutes for 30–40 min (*see* **Note 4**). Vortex for 10 s every 10 min.
6. Stop digestion by adding 1/5 volume of FBS. Mix well by pipetting. Pour the solution through a100 μm cell strainer into a new 50 mL sterile tube.
7. Centrifuge at room temperature at 600 × *g* for 5 min.
8. Remove the supernatant, and resuspend pellet in appropriate volume of complete culture medium. A general rule is 1 mL culture medium for one mouse.
9. Plate the cells into culture plates (6–24 well) and place the plates in a 37 °C 5 % CO_2 Incubator.
10. Next day: Wash cells with pre-warmed PBS to remove the floating cells and add fresh primary brown preadipocytes culture medium. Change the medium every 2 days (*see* **Note 5**).

3.2 Differentiate Primary Brown Preadipocytes

1. When preadipocytes grow to 100 % confluence, change medium with induction medium (day 0) (*see* **Notes 6** and 7).
2. After 48 h (day 2), change induction medium with maintain medium.
3. Change medium every 2 days. Oil droplets will appear around 2–3 days after adding the induction medium. Preadipocytes will differentiate to mature adipocytes around 6–7 days after the induction (Fig. 2).

4 Notes

1. Mouse age: We prefer the new born mice for preadipocyte isolation, although it is also possible to use young mice at 4–5 weeks of age. In our hands, the differentiation is better and UCP1 expression is higher when new born mice are used.

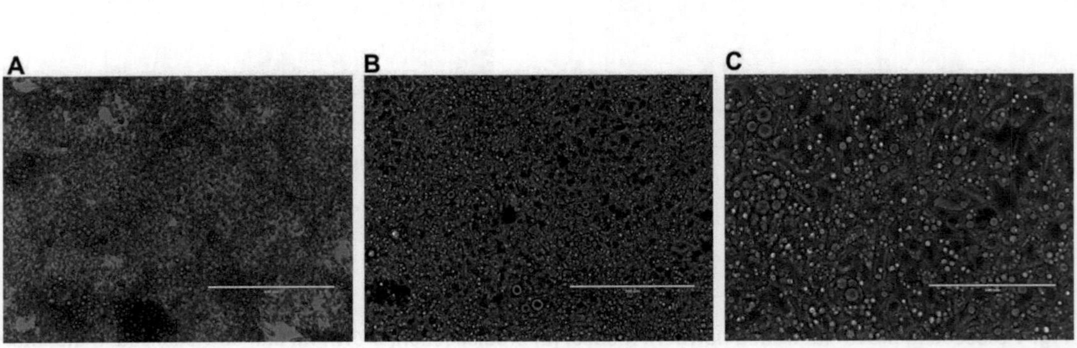

Fig. 2 Differentiated brown adipocytes 6 days after the induction. (**a**) 10×; bar, 400 μm. (**b**) 20×; bar, 200 μm. (**c**) 40×; bar, 100 μm

The UCP1 cycles in qPCR are typically 19–21 for the differentiated brown adipocytes from new born mice.

2. Yield of preadipocytes: We usually use 10–12 new born (1–2 days old) mice for a 12-well plate. Fewer mice can be used if the cells are split for passaging (see below "Cell passaging").
3. Collagenase: There can be significant lot-to-lot variations in collagenase activity. It is advisable to perform a pilot experiment to test the specific lot of collagenase. Once satisfactory results are obtained, a bulky order for that specific lot should be placed. The collagenase can be stored in 4 °C for many years.
4. Digestion time: The digestion time is critical for the successful isolation of preadipocytes. Digestion should not exceed 40 min. The tissue needs to be well digested (the digestion buffer becomes turbid without visible red-colored tissue), but excessive digestion (the digestion buffer turns to be clear again) can damage the cells.
5. Cell passaging: Differentiation tends to decrease quite significantly with passaging. When we have enough preadipocytes to start with, we usually try to avoid passaging. In case that limited new born mice are available, it is possible to passage the cells once, even twice. However, notable differences in adipocyte differentiation rate can be observed after passaging.
6. For differentiation, the preadipocytes must reach 100 % confluence before the induction of differentiation. It is suggested the induction medium needs to be prepared fresh each time.
7. FBS lots may significantly affect adipocyte differentiation [22]. It is advisable to test several lots and purchase in bulk the specific lot with greatest efficiency of inducing adipocyte differentiation.

References

1. Hill JO, Wyatt HR, Peters JC (2012) Energy balance and obesity. Circulation 126:126–132
2. Kajimura S, Spiegelman BM, Seale P (2015) Brown and beige fat: physiological roles beyond heat generation. Cell Metab 22:546–559
3. Rosen ED, Spiegelman BM (2014) What we talk about when we talk about fat. Cell 156:20–44
4. Schulz TJ, Tseng YH (2013) Brown adipose tissue: development, metabolism and beyond. Biochem J 453:167–178
5. Feldmann HM, Golozoubova V, Cannon B, Nedergaard J (2009) UCP1 ablation induces obesity and abolishes diet-induced thermogenesis in mice exempt from thermal stress by living at thermoneutrality. Cell Metab 9:203–209
6. Gospodarska E, Nowialis P, Kozak LP (2015) Mitochondrial turnover: a phenotype distinguishing brown adipocytes from interscapular brown adipose tissue and white adipose tissue. J Biol Chem 290:8243–8255
7. Nedergaard J, Golozoubova V, Matthias A, Asadi A, Jacobsson A, Cannon B (2001) UCP1: the only protein able to mediate adaptive non-shivering thermogenesis and metabolic inefficiency. Biochim Biophys Acta 1504:82–106
8. Betz MJ, Enerback S (2015) Human brown adipose tissue: what we have learned so far. Diabetes 64:2352–2360

9. Schulz TJ, Tseng YH (2013) Brown adipose tissue: development, metabolism and beyond. Biochem J 453:167–178
10. Cypess AM, Haft CR, Laughlin MR, Hu HH (2014) Brown fat in humans: consensus points and experimental guidelines. Cell Metab 20:408–415
11. Cypess AM, Lehman S, Williams G, Tal I, Rodman D, Goldfine AB, Kuo FC, Palmer EL, Tseng YH, Doria A, Kolodny GM, Kahn CR (2009) Identification and importance of brown adipose tissue in adult humans. N Engl J Med 360:1509–1517
12. Nedergaard J, Bengtsson T, Cannon B (2010) Three years with adult human brown adipose tissue. Ann N Y Acad Sci 1212:E20–E36
13. van Marken Lichtenbelt WD, Vanhommerig JW, Smulders NM, Drossaerts JM, Kemerink GJ, Bouvy ND, Schrauwen P, Teule GJ (2009) Cold-activated brown adipose tissue in healthy men. N Engl J Med 360:1500–1508
14. Yoneshiro T, Aita S, Matsushita M, Kayahara T, Kameya T, Kawai Y, Iwanaga T, Saito M (2013) Recruited brown adipose tissue as an antiobesity agent in humans. J Clin Invest 123:3404–3408
15. Cypess AM, Weiner LS, Roberts-Toler C, Franquet Elia E, Kessler SH, Kahn PA, English J, Chatman K, Trauger SA, Doria A, Kolodny GM (2015) Activation of human brown adipose tissue by a beta3-adrenergic receptor agonist. Cell Metab 21:33–38
16. Aune UL, Ruiz L, Kajimura S (2013) Isolation and differentiation of stromal vascular cells to beige/brite cells. J Vis Exp 73:50191
17. Cannon B, Nedergaard J (2001) Cultures of adipose precursor cells from brown adipose tissue and of clonal brown-adipocyte-like cell lines. Methods Mol Biol 155:213–224
18. Fasshauer M, Klein J, Kriauciunas KM, Ueki K, Benito M, Kahn CR (2001) Essential role of insulin receptor substrate 1 in differentiation of brown adipocytes. Mol Cell Biol 21:319–329
19. Gentile P, Orlandi A, Scioli MG, Di Pasquali C, Bocchini I, Cervelli V (2012) Concise review: adipose-derived stromal vascular fraction cells and platelet-rich plasma: basic and clinical implications for tissue engineering therapies in regenerative surgery. Stem Cells Transl Med 1:230–236
20. Shinoda K, Luijten IH, Hasegawa Y, Hong H, Sonne SB, Kim M, Xue R, Chondronikola M, Cypess AM, Tseng YH, Nedergaard J, Sidossis LS, Kajimura S (2015) Genetic and functional characterization of clonally derived adult human brown adipocytes. Nat Med 21:389–394
21. Xue R, Lynes MD, Dreyfuss JM, Shamsi F, Schulz TJ, Zhang H, Huang TL, Townsend KL, Li Y, Takahashi H, Weiner LS, White AP, Lynes MS, Rubin LL, Goodyear LJ, Cypess AM, Tseng YH. (2015) Clonal analyses and gene profiling identify genetic biomarkers of the thermogenic potential of human brown and white preadipocytes, Nat Med 21:767–868.
22. Lee MJ, Fried SK (2014) Optimal protocol for the differentiation and metabolic analysis of human adipose stromal cells. Methods Enzymol 538:49–65.

Chapter 2

Isolation of Mouse Stromal Vascular Cells for Monolayer Culture

Longhua Liu, Louise D. Zheng, Sarah R. Donnelly, Margo P. Emont, Jun Wu, and Zhiyong Cheng

Abstract

Positive energy balance contributes to adipose tissue expansion and dysfunction, which accounts largely for obesity and related metabolic disorders. Thermogenic fat can dissipate energy, activation or induction of which may promote energy balance and address the pressing health issues. Recent studies have shown that stromal vascular fraction (SVF) from white adipose tissue (WAT) can develop both white and brown-like adipocyte phenotypes, thus serving as a unique model to study adipogenesis and thermogenesis. Here, we describe a protocol for effective isolation of mouse SVF from WAT, induction of differentiation, and detection of adipogenesis. Success tips for isolation and culture of SVF are also discussed.

Key words Adipose tissue, Stromal vascular fraction, Adipogenesis, Lipid accumulation, Energy balance, Obesity

1 Introduction

Obesity is one of the most pressing health issues worldwide [1, 2]. Imbalance between energy intake and expenditure contributes to aberrant adiposity, which is associated with changes in mitochondrial function, hormonal signaling, redox status, and inflammatory responses [3–10]. These changes account largely for obesity-related medical conditions including diabetes, cardiovascular diseases, and cancer [2, 11, 12]. The mechanism of obesity-induced metabolic disorders is complex, but it has been increasingly recognized that adipose tissue dysfunction impairs energy expenditure (brown adipose tissue, BAT), energy storage (white adipose tissue, WAT), and endocrine regulation of metabolism via adipokines [3–6]. Thus, targeting adipocyte tissue function to promote thermogenesis for energy homeostasis, such as activation of BAT and browning of WAT, has attracted much interest [13–16].

Jun Wu (ed.), *Thermogenic Fat: Methods and Protocols*, Methods in Molecular Biology, vol. 1566,
DOI 10.1007/978-1-4939-6820-6_2, © Springer Science+Business Media LLC 2017

Adipose tissue is composed of other cell types (collectively called stromal vascular fraction, SVF) in addition to adipocytes, including precursor stem cells, preadipocytes, and fibroblasts [17, 18]. Although immortalized cell lines (e.g., 3T3-L1, 3T3-F442A) provide convenient models for adipogenic study, differences in WAT marker expression and trans-differentiation potential have been recognized between these cell lines and primary cells isolated from adipose tissues [17–19]. For instance, SVF isolated from WAT can develop both white and brown-like adipocyte (brite or beige cell) phenotypes upon cold or adrenergic stimulation, and a bidirectional interconversion exists [20–22]. Therefore, isolation and use of SVF from adipose tissues is of critical importance to understand adipocyte biology. In this chapter, we provide a protocol that demonstrates effective isolation of SVF for monolayer culture and adipogenic study.

2 Materials

2.1 Isolation of SVF

1. Equipment and supplies: 50-mL sterile conical tubes, 0.2-μm filter units, scissors and forceps (autoclave to sterilize), cell strainer (40 μm), light microscope, shaking water bath, bench-top centrifuge, and biosafety cabinet.
2. Krebs–Ringer bicarbonate (KRB) buffer: 118 mM NaCl, 5 mM KCl, 2.5 mM $CaCl_2$, 2 mM KH_2PO_4, 2 mM $MgSO_4$, 25 mM $NaHCO_3$, 5 mM glucose, pH 7.4. KRB buffer is sterilized using 0.2-μm filter units.
3. Phosphate-buffered saline.
4. Collagenase solution: 300 U/mL collagenase (type I) in sterile KRB, with 1 % bovine serum albumin (fraction V).
5. Basal medium: DMEM/F12, 10 % Fetal Bovine Serum (FBS), and 1× Pen/Strep.
6. Mice: 8–12 week old, male C57BL6/J mice were housed as described previously, in plastic cages on a 12-h light–dark photocycle and with free access to water and regular chow diet [1]. Animal use procedures followed the National Institutes of Health guidelines and were approved by the Virginia Tech Institutional Animal Care and Use Committee.

2.2 SVF Monolayer Culture

1. Equipment and supplies: cell culture dishes and plates (collagen coated and noncoated), 0.2-μm filter units, inverted microscope, laboratory CO_2 water-jacketed incubators, bench-top centrifuge, and biosafety cabinet.
2. Phosphate-buffered saline.
3. 0.25 % Trypsin in HBSS.
4. Basal medium: DMEM/F12, 10 % FBS, and 1× Pen/Strep.

2.3 Induction of Adipogenesis

1. Equipment and supplies: inverted microscope, laboratory CO_2 water-jacketed incubators, and biosafety cabinet.
2. Differentiation medium: DMEM/F12 medium containing 10 % FBS, 1× Pen/Strep, dexamethasone (5 μM), insulin (0.5 μg/mL), IBMX (0.5 mM), and rosiglitazone (1 μM).
3. Maintenance medium: DMEM/F12 medium containing 10 % FBS,1× Pen/Strep, and insulin (0.5 μg/mL).

2.4 Detection of Adipogenesis

1. Equipment and supplies: 0.2-μm filter units, 96-well plates, inverted microscope, microplate reader, and western blot analysis system.
2. Oil Red O.
3. Isopropanol.
4. 10 % neutral buffered formalin.
5. Phosphate-buffered saline.
6. Chemiluminescent horseradish peroxidase (HRP) substrates.
7. Primary antibodies against adiponectin.
8. Primary antibodies against PPARγ.
9. Primary antibodies against GAPDH.
10. Secondary antibodies conjugated to HRP.
11. PLC lysis buffer: 30 mM Hepes, pH 7.5, 150 mM NaCl, 10 % glycerol, 1 % Triton X–100, 1.5 mM $MgCl_2$, 1 mM EGTA, 10 mM NaPPi, 100 mM NaF, 1 mM Na_3VO_4, supplemented with protease inhibitor cocktail and 1 mM PMSF immediately before use [1, 23].

3 Methods

3.1 Isolation of SVF

1. Sacrifice the mice by CO_2 asphyxiation, followed by cervical dislocation as a second means of euthanasia.
2. Apply 75 % ethanol spray to the fur for sterilization.
3. To collect subcutaneous (inguinal) white adipose tissue (sWAT), remove the skin along the thighs to reveal sWAT pads on both sides; carefully dissect sWAT pads, and quickly remove any contamination (e.g., hair, muscle, or connective tissue). Store sWAT pads in sterile KRB buffer (on ice) briefly while additional samples are collected.
4. To collect visceral (epidydimal) white adipose tissue (eWAT), change gloves and use another set of clean sterile surgery tools. Cut the peritoneum to expose abdominal organs, locate the testes, and identify the attached eWAT. Carefully dissect eWAT pads from both sides, and quickly remove any contamination (e.g., hair, testes, epididymides, and vasa deferentia). Store the

eWAT pads in sterile KRB buffer (on ice) briefly while additional samples are collected.

5. Quickly dry sWAT and eWAT pads on Kim Wipe paper to remove residual KRB, and place the tissues on clean dry petri dishes.
6. Add $CaCl_2$ (at a final concentration of 2 mM) to freshly made collagenase solution to make digestion medium.
7. Add 10 mL digestion medium to the clean dry petri dishes where sWAT and eWAT are placed, mince the tissues into small pieces (1–2 mm).
8. Cut 2–3 mm from the end of 5 mL pipettes, and transfer the minced tissues with digestion medium into 50 mL conical tubes. Mix very well by pipetting up and down.
9. Incubate the tubes in a 37 °C water bath with constant agitation at 75 rpm for 40–50 min. Check every 10 min to make sure the digestion is sufficient but not excessive (*see* **Note 1**).
10. Stop digestion by adding 10 mL basal medium to the digested tissue homogenate and pipetting to neutralize the collagenase.
11. Centrifuge at 500 × *g* for 5 min; remove the tubes and shake vigorously for 5 s, and centrifuge at 500 × *g* for 5 min.
12. Carefully aspirate the liquid layer and the primary mature adipocyte located on the top, and SVF appears as a brownish pellet on the bottom of the tube.
13. Add 10 mL basal medium to the pellet to resuspend the cells, and centrifuge at 500 × *g* for 5 min.
14. In the biosafety cabinet, carefully aspirate the medium without disturbing the pellet (*see* **Note 2**).
15. Resuspend the pellet in 10 mL basal medium, and filter the cell suspension into a new tube using a cell strainer (40 μm diameter).
16. Centrifuge at 500 × *g* for 5 min, and carefully aspirate the medium without disturbing the pellet.

3.2 SVF Monolayer Culture

1. Resuspend the pellet from Subheading 3.1, **step 16**, in 10 mL pre-warmed basal medium.
2. Plate the cells evenly onto a 10-cm collagen-coated plate, and place in a CO_2 incubator (37 °C, 5 % CO_2) (*see* **Note 3**).
3. Gently aspirate medium 2 h after cell plating, and wash the cells three times with PBS (*see* **Note 4**).
4. Add 10 mL fresh pre-warmed basal medium to each dish, and place back into the CO_2 incubator (37 °C, 5 % CO_2). Change medium every 2 days.

5. At a 95 % confluence, split cells for subculture from one 10-cm dish into three 10-cm dishes, three 6-well plates, or three 12-well plates for later differentiation (*see* **Note 5**). Change medium every 2 days.

3.3 Induction of Adipogenesis

1. When the cell subcultures grow to a 95 % confluence for eWAT or 2 days after confluence for sWAT (day 0), replace the basal medium with differentiation medium. Supply fresh differentiation medium every 2 days.
2. Four days later (day 4), replace the differentiation medium with maintenance medium. Supply fresh maintenance medium every 2 days. Development of lipid droplets in the cells becomes obvious within 4 days after the addition of differentiation media.
3. Four days later (day 8), lipid accumulation is complete, and adipocytes can be harvested for protein analysis and Oil Red O staining (*see* **Note 6**).

3.4 Detection of Adipogenesis

1. Dissolve 0.5 g Oil Red O in 100 mL isopropanol, and filter through a 0.2-μm filter.
2. Immediately before staining experiments, add 30 mL of the Oil Red O stock to 20 mL distilled water to make Oil Red O working solution. Use this solution within 24 h of its preparation, and filter again if precipitation is present.
3. Remove maintenance medium and wash the cells three times with PBS.
4. Add 10 % buffered formalin (i.e., 4 % formaldehyde) sufficient to completely cover the cells and let sit for 10 min at room temperature.
5. Remove fixative, and rinse cells with PBS for 1 min, followed by two washes with water for 1 min.
6. Let the cell monolayers air dry for 10 min.
7. Stain the cells with Oil Red O solution (0.5 % in 60 % isopropanol) for 30 min, and rinse with water 3 times.
8. Visualize the stained cells on an inverted microscope system (Fig. 1).
9. Extract Oil Red O from the cells using 100 % isopropanol, gentle agitation for 15 min at room temperature.
10. Transfer the extracts to a 96-well plate (100 μl per well), and quantify lipid accumulation in the differentiated cells by measuring absorbance at 510 nm on a microplate reader (Fig. 1).
11. Additional detection of adipogenesis is performed by western blot analysis of PPARγ (a master adipogenic regulator) and adiponectin (an adipokine secreted by differentiated adipocytes)

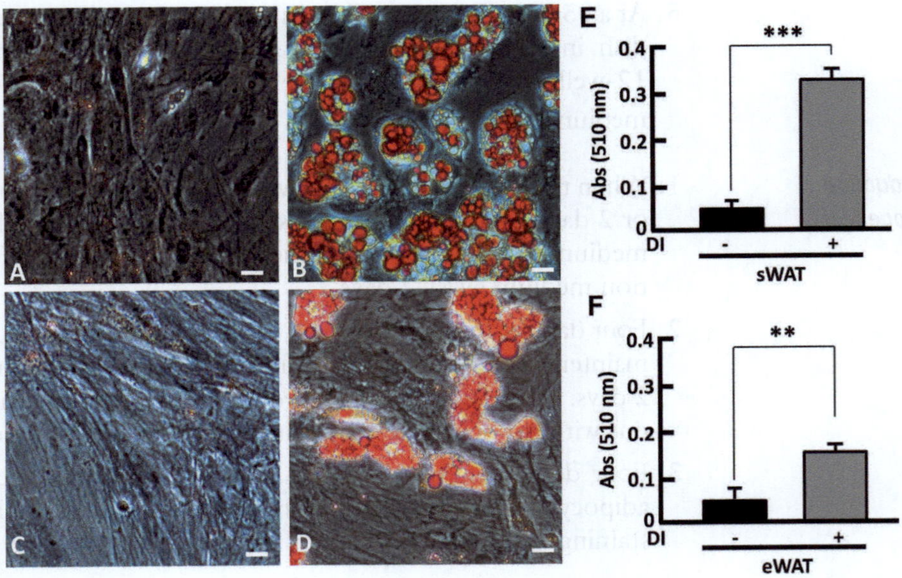

Fig. 1 Detection of SVF differentiation by Oil Red O staining. (**a**) Staining and imaging of SVF monolayers isolated from sWAT at day 8 without (**a**) and with (**b**) differentiation induction. (**c–d**) Staining and imaging of SVF monolayers isolated from eWAT at day 8 without (**c**) and with (**d**) differentiation induction. (**e**) Lipid accumulation in SVF from sWAT was measured by absorbance at 510 nm on a microplate reader. (**f**) Lipid accumulation in SVF from eWAT was measured by absorbance at 510 nm on a microplate reader. Scale bars, 50 μm. DI, differentiation induction; $n = 3–5$; **, $p < 0.01$; ***, $p < 0.0001$

using the standard procedure as described previously [1, 23], which shows upregulation of PPARγ and adiponectin (Fig. 2). In line with the pattern of lipid accumulation (Fig. 1), PPARγ and adiponectin were upregulated to a greater extent in SVF isolated from sWAT than from eWAT after differentiation induction.

4 Notes

1. If a 37 °C water bath is unavailable, the tissue digestion can be performed in an incubator shaker with setting at 150 rpm, 37 °C. Typically, a well-digested tissue appears to be a cloudy solution, with no chunks of fat remaining. Over-digestion of tissues will reduce cell yield and viability.
2. Sterile biosafety hood should be used in the following steps.
3. SVF isolated from 2–3 mice show good cell density on a 10-cm dish. A noncoated plate also works, but it takes a longer time (i.e., overnight incubation) for adipogenic cells to fully adhere to the plates. It would reduce adipogenic cell density if the first medium change takes place within 2 h after cell plating in **step 3** in subheading 3.2.

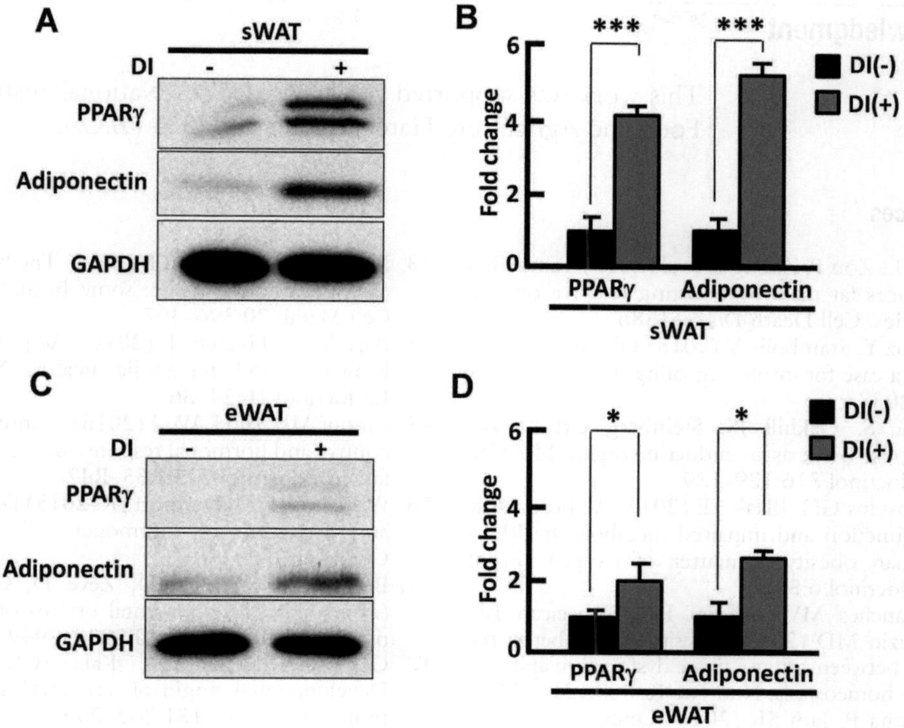

Fig. 2 Detection of SVF differentiation by western blot analysis. (**a**) Representative western blot images of SVF isolated from sWAT at day 8 with or without differentiation induction. (**b**) Densitometric analysis of western blot images acquired as described in panel (**a**). (**c**) Representative western blot images of SVF isolated from eWAT at day 8 with or without differentiation induction. (**d**) Densitometric analysis of western blot images acquired as described in panel (**a**). *DI* differentiation induction; $n = 3–5$; * $p < 0.05$; *** $p < 0.0001$

4. This washing step is critical to remove contaminated red blood cells, immune cells, and other contaminants. Alternatively, ACK lysis buffer (154 mM NH_4Cl, 10 mM $KHCO_3$ and 0.1 mM EDTA, pH 7.4) can be used to remove red blood cells by resuspending the SVF pellet and incubated for 5 min at room temperature. After centrifuging at 500 × *g* for 5 min, remove the ACK buffer and wash SVF with basal medium once, then continue with **step 15** in Subheading 3.1.
5. The primary stromal vascular cells grow much slower when they reach passage 3 or higher. We use cells at passage 2 for differentiation experiments. No difference is observed between collagen coated and non-coated plates during the cell subculture and adipogenic study.
6. SVF isolated from sWAT has a greater differentiation potential than that from eWAT, presumably because the latter requires a 3D structural support to stimulate its intrinsic differentiation potential. Indeed, SVF from eWAT pads exhibits robust differentiation in a 3D hydrogel system [24].

Acknowledgment

This work was supported in part by USDA National Institute of Food and Agriculture Hatch Project 1007334 (Z.C.).

References

1. Liu L, Zou P, Zheng L et al (2015) Tamoxifen reduces fat mass by boosting reactive oxygen species. Cell Death Dis 6:e1586
2. Poloz Y, Stambolic V (2015) Obesity and cancer, a case for insulin signaling. Cell Death Dis 6:e2037
3. Galic S, Oakhill JS, Steinberg GR (2010) Adipose tissue as an endocrine organ. Mol Cell Endocrinol 316:129–139
4. Goossens GH, Blaak EE (2015) Adipose tissue dysfunction and impaired metabolic health in human obesity: a matter of oxygen? Front Endocrinol 6:55
5. Abranches MV, Oliveira FC, Conceicao LL, Peluzio MD (2015) Obesity and diabetes: the link between adipose tissue dysfunction and glucose homeostasis. Nutr Res Rev 28:121–132
6. Manna P, Jain SK (2015) Obesity, oxidative stress, adipose tissue dysfunction, and the associated health risks: causes and therapeutic strategies. Metab Syndr Relat Disord 13:423–444
7. Saltiel AR (2012) Insulin resistance in the defense against obesity. Cell Metab 15:798–804
8. Hill JO, Wyatt HR, Peters JC (2012) Energy balance and obesity. Circulation 126:126–132
9. Cheng Z, Almeida FA (2014) Mitochondrial alteration in type 2 diabetes and obesity: an epigenetic link. Cell Cycle 13:890–897
10. Cummins TD, Holden CR, Sansbury BE et al (2014) Metabolic remodeling of white adipose tissue in obesity. Am J Physiol Endocrinol Metab 307:E262–E277
11. van der Klaauw AA, Farooqi IS (2015) The hunger genes: pathways to obesity. Cell 161:119–132
12. Cheng Z, Ristow M (2013) Mitochondria and metabolic homeostasis. Antioxid Redox Signal 19:240–242
13. Nedergaard J, Cannon B (2014) The browning of white adipose tissue: some burning issues. Cell Metab 20:396–407
14. Bartelt A, Heeren J (2014) Adipose tissue browning and metabolic health. Nat Rev Endocrinol 10:24–36
15. Emont MP, Yu H, Wu J (2015) Transcriptional control and hormonal response of thermogenic fat. J Endocrinol 225:R35–R47
16. Wu J, Jun H, McDermott JR (2015) Formation and activation of thermogenic fat. Trends Genet 31:232–238
17. Berry DC, Stenesen D, Zeve D, Graff JM (2013) The developmental origins of adipose tissue. Development 140:3939–3949
18. Gesta S, Tseng YH, Kahn CR (2007) Developmental origin of fat: tracking obesity to its source. Cell 131:242–256
19. Lafontan M (2012) Historical perspectives in fat cell biology: the fat cell as a model for the investigation of hormonal and metabolic pathways. Am J Physiol Cell Physiol 302:C327–C359
20. Lee YH, Petkova AP, Mottillo EP, Granneman JG (2012) In vivo identification of bipotential adipocyte progenitors recruited by beta3-adrenoceptor activation and high-fat feeding. Cell Metab 15:480–491
21. Rosenwald M, Perdikari A, Rulicke T, Wolfrum C (2013) Bi-directional interconversion of brite and white adipocytes. Nat Cell Biol 15:659–667
22. Wu J, Bostrom P, Sparks LM et al (2012) Beige adipocytes are a distinct type of thermogenic fat cell in mouse and human. Cell 150:366–376
23. Zou P, Liu L, Zheng L et al (2014) Targeting FoxO1 with AS1842856 suppresses adipogenesis. Cell Cycle 13:3759–3767
24. Emont MP, Yu H, Jun H et al (2015) Using a 3D culture system to differentiate visceral adipocytes in vitro. Endocrinology 156:4761–4768

Chapter 3

Flow Cytometry Assisted Isolation of Adipose Tissue Derived Stem Cells

Umesh D. Wankhade and Sushil G. Rane

Abstract

Adipose tissue dysfunction is typically seen in metabolic diseases, particularly obesity and diabetes. White adipocytes store fat while brown adipocyte dissipates it via thermogenesis. In addition, beige adipocytes develop in white fat depots in response to stimulation of β-adrenergic pathways. It appears that the three types of adipocytes—white, brown, and beige—can be formed de novo from stem/precursor cells or via transdifferentiation. Identifying the presumptive progenitors that harbor capacity to differentiate to these distinct adipocyte cell types will enable their functional characterization. Moreover, the presence or absence of white/brown/beige adipocytes is correlated with metabolic dysfunction making their study of medical relevance. Robust, reliable, and reproducible methods of identification and isolation of adipocyte progenitors will stimulate further detailed understanding of white, brown, and beige adipogenesis.

Key words Obesity, Adipose tissue, WAT, BAT, Beige adipose, Stem cells, Progenitor cells, ADSCs, Flow cytometry

1 Introduction

Adipose tissue dysfunction, along with insulin resistance and pancreatic β-cell failure, is at the core of the obesity and diabetes epidemic [1–3]. White adipose tissue (WAT) depots are dispersed in distinct locations and serve the prime function of fat storage. In contrast, brown adipose tissue (BAT), typically localized to the supraclavicular and neck regions and along the spine, dissipates fat via a process termed thermogenesis [4], driven primarily by the inner mitochondrial membrane protein Ucp1. Recent findings that metabolically active brown adipose tissue (BAT) exists in adult humans [5, 6], have renewed widespread interest in its therapeutic potential to combat metabolic diseases [7, 8]. In addition to the classical BAT, adipocytes expressing variable levels of Ucp1—termed beige or brite adipocytes—appear in white fat depots in response to cold exposure or upon stimulation by β-adrenergic pathways [9, 10].

Jun Wu (ed.), *Thermogenic Fat: Methods and Protocols*, Methods in Molecular Biology, vol. 1566, DOI 10.1007/978-1-4939-6820-6_3, © Springer Science+Business Media LLC 2017

A summarization of several excellent studies suggests that brown and beige adipocytes may have a distinct developmental origin [11, 12] with evidence favoring the existence of specialized progenitors that drive their genesis [13–15]. In addition to de novo adipogenesis, a role for trans-differentiation has been proposed as a mechanism underlying beige/brite adipogenesis [16]. Further, a bi-potential progenitor that differentiates toward white or brown adipocytes has been identified [17].Notably, Sca-1+/CD45-/Mac1- cells [14] and PDGFRα+/CD34+/Sca-1+ cells [17] have been suggested to have brown adipogenic potential, although it is unclear if these presumptive progenitors also promote beige adipocyte differentiation. Wang et al. showed that the rate of white adipogenesis in both epididymal and subcutaneous adipose tissue varies with diet. Early exposure to HFD leads to adipose tissue expansion because of hypertrophy of adipocytes whereas more chronic exposure to HFD results in extensive adipogenesis of gonadal fat tissue, but not subcutaneous adipose tissue [15]. With lineage tracking experimentation, Wang et al. demonstrated that most of the newly emerging beige adipocytes in subcutaneous depots were not derived from preexisting white adipocytes, thus suggestive of a potential precursor cell that undergoes differentiation [15]. Wu et al. demonstrated that a subset of precursor cells within subcutaneous adipose tissue can give rise to beige cells, which are capable of expressing abundant UCP1 and a broad gene expression program that is distinct from either white or classical brown adipocytes. These inducible beige progenitor cells were sorted by flow cytometry based on expression of beige-selective cell surface proteins CD137 or TMEM26 [13].

Taken together, these data define a population of tissue resident, inducible beige/brown-adipocyte progenitors in mice. Understanding beige/brown adipogenesis, given that the appearance of these cells is associated with improved metabolism in mice [3, 18, 19], is thus of potential clinical relevance for metabolic diseases. Particularly, reliable and reproducible methods to identify, isolate, and characterize presumptive progenitors that retain capacity to differentiate to the three adipocyte lineages—white, brown, and beige—are important to further our understanding of their biology and functionality.

2 Materials

2.1 Tissue

Adipose tissue from several depots can be used to isolate ADSCs. We used epididymal/gonadal adipose tissue depot.

2.2 Supplies

1. Beakers.
2. Mincing scissors.

3. Scalpel.
4. CO_2.
5. 50 mL conical tubes.
6. 15 mL conical tubes.
7. 1 mL microfuge tubes.
8. 100 um mesh filter.
9. 5 mL round-bottom tubes for FACS.
10. 2.5 cm cell culture dishes.
11. Hemocytometer.
12. CO_2 gas chamber.
13. Water bath with shaker.
14. Benchtop Centrifuge.
15. Biosafety/Cell Culture hood.
16. Vacuum filtration assembly.
17. Microscope.
18. CO_2 incubator.

2.3 Buffers and Culturing Media

1. *Adipose tissue digestion buffer (a),* NaCl (123 mM); KCl (5 mM); $CaCl_2$ (1.3 mM); Glucose (5 mM); Hepes (100 mM); BSA (4 %); Collagenase I (1 mg/mL); Water to 50 mL.
2. *Collagenase solution:* Weigh out 0.1 g of type I collagenase and dissolve it in 1 mL adipocyte isolation buffer. This solution can be stored for longer period at −20 °C.
3. *FACS Buffer,* PBS with Mg^{++} and Ca^{++} (1×); EDTA (1 mM); Hepes (25 mM); Fatty Acid Free BSA (1 %); Dissolve bovine serum albumin (fraction V), EDTA, and HEPES in 500 mL of phosphate buffered saline (PBS). After sterile filtration, warm the solution to 37 °C. This solution should be used within 1 h of its preparation.
4. *Culture media:* To 500 mL of DMEM media, add 110 mL of fetal bovine serum (20 %) and 5.6 mL of antibiotic (penicillin/streptomycin)/antimycotic 100× stock solutions. All the media solutions are filtered through a 0.2-μm filter unit.

2.4 Antibodies

1. CD31, 1:100, (Fluroschrome BV421).
2. CD34, 1:100, (Fluroschrome Pe-Cy5).
3. CD45, 1:100, (Fluroschrome APC-Cy7).
4. CD146, 1:100, (Fluroschrome PerCP-Cy5.5).
5. CD90, 1:100, (Flurochrome PE).
6. CD105, 1:100, (Flurochrome AF 647).
7. Live/Dead Aqua Antibody, 1:500.

3 Methods

3.1 Adipose Tissue Depot Excision and Preparation

1. Euthanize mice with CO_2 exposure at the rate of 3 L/min. Continue CO_2 until 1 min after breathing stops. Confirm euthanasia by performing cervical dislocation (*see* **Note 1**).
2. Immerse the whole mouse in 70 % ethanol for 2–3 min followed by wash in PBS (*see* **Note 2**).
3. Pin the mouse to the dissecting surface. Open the mouse and remove the epididymal/gonadal adipose tissue depot. Make sure not to mix the adipose tissue from different depots if investigating depot specific differences (*see* **Note 3**).
4. Collect adipose tissue in 50 mL conical tube containing adipose tissue digestion buffer. If you are extracting adipose tissues from multiple mice then keep tube on ice until you are ready to mince the tissue (*see* **Note 4**).
5. Start with 5 mL per fat pad digestion buffer for lean mice and 10 mL per fat pad for obese mice (*see* **Note 5**).
6. Mince tissue into small pieces in up to 9 mL (1 mL/0.1 g tissue) supplemented digestion buffer.
7. Use scissor to mince the tissue in fine pieces. It usually takes 2–3 min to mince adipose tissue to <1 mm pieces (Fig. 1).

3.2 Digestion and Incubation

1. Vortex tube containing buffer with finely minced pieces of adipose tissue and incubate at 37 °C in water bath with shaker speed @ 220–250 RPM for 45–60 min.
2. Vortex the tube every 10 min and make sure tissue pieces do not stick to wall of tube.
3. After complete digestion, samples are poured onto a 100 μm nylon mesh into a new 50 mL tube to remove noncellular fibrous material and undigested tissue. Rinse the filter with ~5 mL of supplemented digestion buffer (*see* **Note 6**).
4. Spin the filtrate at 250 × *g* for 5 min.
5. Remove the floating adipocyte layer (put on dry ice ASAP if using for RNA/protein) and most of supernatant with a (transfer) pipet into a new 15 mL tube and spin again. Collect the 1st pellet as the stromal vascular cell fraction (SVF) (*see* **Note 7**) (Fig. 1).

3.3 Purification and Preparation of SVF for FACS

1. SVF portion from the last step consists of several types of cells such as monocytes, lymphocytes, RBCs, and others. To eliminate red blood cells use ACK lysis buffer.
2. Incubate the SVF with 0.5 or 1.0 mL of ACK lysis buffer for 1 min at room temperature.
3. Wash the pellet with 5 mL of FACS buffer and spin at 500 × *g* for 10 min.

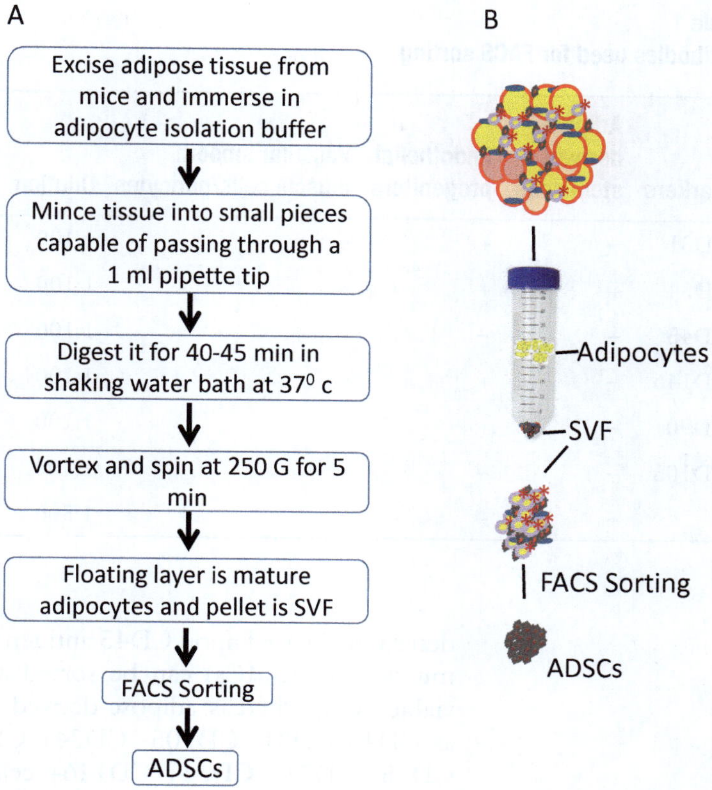

Fig. 1 (**a**) Step by step flow chart of protocol, (**b**) Schematic representation of protocol

4. Centrifuge at 500 × *g* for 5 min and resuspend pellet in chilled sorting buffer to a concentration of 10^6 cells/100 μL.
5. Freeze extra SVCs for RNA/protein extraction according to your experimental needs @ −80 °C (Fig. 1).

3.4 Staining and Sorting

1. Get two Ice buckets and foil to keep cells from light exposure.
2. Resuspend cells @ 1 million/100 μL in cold sorting buffer.
3. Refer to the antibody dilutions information in Table 1. Add the antibodies one by one to the FACS buffer containing pellet. Mix well by pipetting. Incubate on ice for 30–45 min. Stained samples are washed twice and later sorted on FACS Aria sorter (BD Biosciences, USA) equipped with 407, 488, 532, and 633 LASER lines using DIVA v6.1.3 software.

3.5 Gating and Selection Strategy for Sorting

Populations are identified and sorted as per the gating strategy desired. This can be modified to what is needed per the experimental need. Briefly, after excluding cellular debris using a forward light scatter/side scatter plot, viable CD45 negative gate was

Table 1
Antibodies used for FACS sorting

Markers	Adipose derived stem cells	Endothelial progenitors	Vascular smooth muscle cells/pericytes	Dilution	Fluorochrome
CD31	–	+	–	1:100	BV421 (BD)
CD34	+	+	+	1:100	Pe-Cy5 (BioLegend)
CD45	–	–	–	1:100	APC-Cy7 (BD)
CD146	–	+	+	1:100	PerCP-Cy5.5 (BD)
CD90	+	+	+	1:100	PE (BD)
CD105	–	–	+	1:100	AF 647 (BD)
–	–	–	–	1:500	Live/Dead Aqua (Invitrogen)

determined based upon CD45 antigen staining properties. Smooth muscle cells (SMCs) can be sorted as CD45- CD31- CD105+ viable cells, whereas adipose-derived stem cells (ADSCs) sorted as CD45-CD31- CD105- CD34+ CD90.2+ cells. Furthermore, CD45- CD31+ CD34+ CD146+ cells are sorted as endothelial progenitors.

3.6 Collection and Processing of ADSCs

1. Sorted cells are collected in DMEM media with 20 % FBS. Cells are spun at 100 × *g* for 5 min and washed once with the same media (*see* **Note 8**).
2. Cells are either plated on small diameter cell culture plate or collected for RNA or protein isolation (*see* **Note 9**).
3. ADSCs can subsequently be differentiated to either white or brown adipocytes using established protocols described elsewhere.

4 Notes

1. All animal procedures must conform to the requirements of the Animal Welfare Act and be approved before implementation by the Institutional Animal Care and Use Committee (IACUC).
2. To maintain the sterile conditions during cell culture, frequent spraying of surface area and hands with ethanol is suggested. After euthanasia, mice carcass should be immersed completely in 70 % ethanol for a few minutes before opening up to collect adipose tissue.

3. Because of their proximity to one another care needs to be taken to prevent mixing different adipose tissue depots during excision and isolation of cells.
4. Adipose tissue harvesting should be completed in a quick time-frame and SVC isolation should be initiated within 20–30 min of sacrifice while tissue sitting on ice during waiting period.
5. Stock solutions can be prepared ahead of time, aliquoted, and stored at −80 °C or −20 °C for longer duration. Filter at the time of media preparation. Some filter materials may be sensitive to specific solvents (e.g., DMSO, methanol) and may disintegrate upon exposure; therefore, it is safer to filter stock solutions after they are diluted in the medium.
6. After digestion, adipose tissue can be filtered to remove any remaining undigested pieces. Use a 100-μm filter (BD–Falcon) for filtering small volumes. However, to collect mature adipocytes at a later step, use a nylon filter (sterile) screen with a pore size of 250 μm.
7. In case pieces of undigested tissue in the sample remain after the 1 h digestion, verify that the collagenase solution used is fresh and has not been maintained at room temperature for an extended period of time. This is important to maximize the enzyme efficiency. The collagenase solution can also be stored at −20 °C for a few days, with a minor loss of enzyme activity. Prior to use, the frozen solution can be slowly thawed at room temperature and warmed to 37 °C.
8. High FBS containing media is recommended during initial stages of cell culture.
9. To accelerate cell adhesion, the culture dishes can be precoated with extracellular matrix components, such as gelatin or Matrigel.

References

1. Sun K, Kusminski CM, Scherer PE (2011) Adipose tissue remodeling and obesity. J Clin Invest 121(6):2094–2101. doi:10.1172/JCI45887
2. Gesta S, Tseng YH, Kahn CR (2007) Developmental origin of fat: tracking obesity to its source. Cell 131(2):242–256. doi:10.1016/j.cell.2007.10.004
3. Rosen ED, Spiegelman BM (2014) What we talk about when we talk about fat. Cell 156(1-2):20–44. doi:10.1016/j.cell.2013.12.012
4. Cannon B, Nedergaard J (2004) Brown adipose tissue: function and physiological significance. Physiol Rev 84(1):277–359. doi:10.1152/physrev.00015.2003
5. Nedergaard J, Bengtsson T, Cannon B (2007) Unexpected evidence for active brown adipose tissue in adult humans. Am J Physiol Endocrinol Metab 293(2):E444–E452. doi:10.1152/ajpendo.00691.2006
6. Cypess AM, Lehman S, Williams G, Tal I, Rodman D, Goldfine AB, Kuo FC, Palmer EL, Tseng YH, Doria A, Kolodny GM, Kahn CR (2009) Identification and importance of brown adipose tissue in adult humans. N Engl J Med 360(15):1509–1517. doi:10.1056/NEJMoa0810780
7. Nedergaard J, Cannon B (2010) The changed metabolic world with human brown adipose tissue: therapeutic visions. Cell Metab 11(4):268–272. doi:10.1016/j.cmet.2010.03.007

8. Enerback S (2010) Human brown adipose tissue. Cell Metab 11 (4):248–252.doi:10.1016/j.cmet.2010.03.008,S1550-4131(10)00078-1 [pii]
9. Cousin B, Cinti S, Morroni M, Raimbault S, Ricquier D, Penicaud L, Casteilla L (1992) Occurrence of brown adipocytes in rat white adipose tissue: molecular and morphological characterization. J Cell Sci 103(Pt 4):931–942
10. Guerra C, Koza RA, Yamashita H, Walsh K, Kozak LP (1998) Emergence of brown adipocytes in white fat in mice is under genetic control. Effects on body weight and adiposity. J Clin Invest 102(2):412–420. doi:10.1172/JCI3155
11. Xue B, Rim JS, Hogan JC, Coulter AA, Koza RA, Kozak LP (2007) Genetic variability affects the development of brown adipocytes in white fat but not in interscapular brown fat. J Lipid Res 48(1):41–51
12. Petrovic N, Walden TB, Shabalina IG, Timmons JA, Cannon B, Nedergaard J (2010) Chronic peroxisome proliferator-activated receptor gamma (PPAR gamma) activation of epididymally derived white adipocyte cultures reveals a population of thermogenically competent, UCP1-containing adipocytes molecularly distinct from classic brown adipocytes. J Biol Chem 285(10):7153–7164
13. Wu J, Bostrom P, Sparks LM, Ye L, Choi JH, Giang AH, Khandekar M, Virtanen KA, Nuutila P, Schaart G, Huang K, Tu H, van Marken Lichtenbelt WD, Hoeks J, Enerback S, Schrauwen P, Spiegelman BM (2012) Beige adipocytes are a distinct type of thermogenic fat cell in mouse and human. Cell 150(2):366–376. doi:10.1016/j.cell.2012.05.016
14. Schulz TJ, Huang TL, Tran TT, Zhang H, Townsend KL, Shadrach JL, Cerletti M, McDougall LE, Giorgadze N, Tchkonia T, Schrier D, Falb D, Kirkland JL, Wagers AJ, Tseng YH (2011) Identification of inducible brown adipocyte progenitors residing in skeletal muscle and white fat. Proc Natl Acad Sci U S A 108(1):143–148. doi:10.1073/pnas.1010929108
15. Wang QA, Tao C, Gupta RK, Scherer PE (2013) Tracking adipogenesis during white adipose tissue development, expansion and regeneration. Nat Med 19(10):1338–1344. doi:10.1038/nm.3324
16. Himms-Hagen J, Melnyk A, Zingaretti MC, Ceresi E, Barbatelli G, Cinti S (2000) Multilocular fat cells in WAT of CL-316243-treated rats derive directly from white adipocytes. Am J Physiol 279(3):C670–C681
17. Lee YH, Petkova AP, Mottillo EP, Granneman JG (2012) In vivo identification of bipotential adipocyte progenitors recruited by beta3-adrenoceptor activation and high-fat feeding. Cell Metab 15(4):480–491. doi:10.1016/j.cmet.2012.03.009
18. Bartelt A, Heeren J (2014) Adipose tissue browning and metabolic health. Nat Rev Endocrinol 10(1):24–36. doi:10.1038/nrendo.2013.204
19. Harms M, Seale P (2013) Brown and beige fat: development, function and therapeutic potential. Nat Med 19(10):1252–1263. doi:10.1038/nm.3361

Chapter 4

Flow Cytometric Isolation and Differentiation of Adipogenic Progenitor Cells into Brown and Brite/Beige Adipocytes

Jochen Steinbring, Antonia Graja, Anne-Marie Jank, and Tim J. Schulz

Abstract

Aside from mature adipocytes, adipose tissue harbors several distinct cell populations including immune cells, endothelial cells, and adipogenic progenitor cells (AdPCs). AdPCs represent the reservoir of regenerative cells that replenishes adipocytes during normal cellular turnover and during times of increased demand for triglyceride-storage capacity. The worldwide increase in pathologies associated with the metabolic syndrome, such as obesity and type-2 diabetes, has heightened public and scientific interest in adipose tissues and the cell biological processes of adipose tissue formation and function. Two distinct types of fat cells are known: White and brown adipocytes. Especially brown adipose tissue (BAT) has received considerable attention due to its unique capacity for thermogenic energy expenditure and potential role in the treatment of adiposity. Accordingly, the cold-induced conversion of white into brown-like adipocytes has become a feasible approach in humans and a study-subject in rodents to better understand the underlying molecular processes. Fluorescence-activated cell sorting (FACS) provides a method to isolate AdPCs and other cell populations from adipose tissue by using antibodies detecting unique surface markers. We here describe an approach to isolate cells committed to the adipogenic lineage and summarize established protocols to differentiate FACS-purified primary AdPCs into UCP1-expressing brown adipocytes under in vitro conditions.

Key words Brown/brite adipose tissue, White adipose tissue, Adipogenic progenitor cells, Fluorescence-activated cell sorting, Cell surface marker antibodies, Brown adipogenesis

1 Introduction

At present, the World Health Organization classifies more than 1.9 billion people worldwide as overweight, i.e., with a body mass index (BMI) greater than 25. Among these, approximately 600 million are considered obese with a BMI of 30 or more. Additionally, childhood obesity is a growing concern that likely will have long-lasting health consequences for the future adult population [1]. Traditionally, adipose tissue has been divided into white adipose tissue (WAT), specializing in triglyceride storage, and brown adipose tissue (BAT), which is responsible for

Jun Wu (ed.), *Thermogenic Fat: Methods and Protocols*, Methods in Molecular Biology, vol. 1566,
DOI 10.1007/978-1-4939-6820-6_4,

thermogenesis but mostly restricted to small rodents and newborns with an unfavorable surface-to-volume ratio. Over the past few decades, paralleling the onset of the obesity pandemic, a rising interest in the pathogenesis of the metabolic syndrome has precipitated in significant advances in adipose tissue biology. Inspired by the discovery of leptin in 1994, WAT is now recognized as an important endocrine organ secreting hormonal signaling molecules, the so-called adipokines, that regulate virtually all metabolic processes in the body [2–4].

Similarly, it was not until recently that the presence of BAT in adult humans was rediscovered along with the recognition that it is metabolically active and could affect systemic energy balance [5–10]. Importantly, more recent studies have concomitantly established a direct correlation between BAT activity and weight loss and an improved metabolic profile [11–13]. BAT consists of thermogenic brown adipocytes that are mainly controlled by the sympathetic nervous system (SNS) [14, 15]. It is characterized by its brownish color resulting from a high degree of vascularization and its high mitochondrial density [14]. Thermogenesis is achieved by dissociating the mitochondrial proton gradient from adenosine triphosphate (ATP)-synthesis due to the activity of the uncoupling protein 1 (UCP1). UCP1 is uniquely expressed in brown adipocytes and resides in the inner mitochondrial membrane [14, 15]. Alongside the classical BAT, a second population of adipocytes expressing high levels of UCP1 is recruited within depots of WAT in response to cold-exposure or other means of adrenergic stimulation (reviewed in [16, 17]). Owing to their interspersed occurrence in different depots of white fat, these brown-like adipocytes have also been termed brite (brown-in-white) or beige adipocytes [18–20]. They are developmentally distinct from BAT, which is more closely related to skeletal muscle, but the transcriptional machinery directing their differentiation into brown and beige/brite adipocytes is similar in many aspects [21]. For instance, Seale et al. showed that expression of the transcription factor PR domain containing 16 (PRDM16) induces lineage commitment of embryonic progenitors toward the brown adipogenic lineage and also is required for browning of WAT by influencing the activity of the transcription factors CCAAT/enhancer-binding protein beta (CEBP/β), peroxisome proliferator-activated receptor gamma (PPARγ), peroxisome proliferator-activated receptor alpha (PPARα), and peroxisome proliferator-activated receptor gamma coactivator 1-alpha (PGC1α) [22–24]. Although UCP1 expressing cells are common in all white adipose tissues after prolonged cold exposure in mice, inguinal adipose tissue, a major subcutaneous white fat depot, appears to be the depot that is most receptive to browning-stimuli [25–27]. Accordingly, we were recently able to show that browning of WAT following genetic ablation of the interscapular depot of BAT is sufficient to retain a healthy metabolic profile, suggesting that beige/brite adipose tissue is metabolically equivalent to BAT [28]. Adipose tissues are heterogeneous tissues containing multiple cell types. Macrophages and other immune cells represent up to 10 % of all the cells in adipose tissue of lean mice, while the number increases significantly in obese mice and is accompanied by increased levels of free fatty acids (FFA) and pro-inflammatory cytokines [29]. AdPCs of BAT and WAT express a common set of cell surface markers, namely the platelet-derived growth factor receptor (PDGFR)-α, the stem cell antigen (Sca)-1, and cluster of differentiation (CD)-34 [30–32]. These cells are capable of undergoing adipogenic differentiation in vivo and in vitro. Furthermore, AdPCs are negative for the endothelial cell marker PECAM-1 (CD31) as well as the hematopoietic cell marker proteins leukocyte common antigen (CD45) and integrin αM (CD11b) [31]. In agreement with these observations, lineage tracing studies have also revealed a non-hematopoietic and non-endothelial developmental origin of AdPCs [33, 34]. A specific configuration of cell surface markers, i.e., $Sca1^+$/$PDGFR\alpha^+$/$CD45^-$/$CD31^-$, therefore enables the prospective isolation of highly enriched AdPCs derived from BAT as well as WAT. Exposure to different stimulants, such as rosiglitazone, bone morphogenetic protein (BMP)-7, and prostaglandins among many others, has been described to efficiently induce brown and brite/beige adipocyte differentiation under in vitro conditions to study potential regulatory mechanisms of this process [19, 31, 35, 36].

In this article, we describe a method to isolate AdPCs from inguinal, subcutaneous WAT (scWAT), and BAT by fluorescence associated cell sorting (FACS). To this end, cell surface markers are stained with fluorochrome-labeled antibodies which enables flow cytometric identification and subsequent sorting of specific subsets of cells. This method is not limited to the isolation of AdPCs and can also be applied to isolate and characterize other adipose-tissue resident cell populations. Moreover, we present a protocol to cultivate adipogenic progenitor cells and achieve high levels of brown and brite/beige adipogenic differentiation under in vitro conditions.

2 Materials

2.1 Isolation of Progenitor Cells from Murine Brown and Inguinal White Adipose Tissue

1. Sterile dissection instruments.
2. Fetal bovine serum (FBS).
3. Hanks' balanced salt solution (HBSS).
4. Bovine serum albumin (BSA).
5. Water bath.
6. 50 mL conical reaction tubes.

7. ACK lysis buffer: 0.15 M ammonium chloride, 0.01 M potassium bicarbonate, 0.01 M ethylenediaminetetraacetic acid (EDTA) in 1 L distilled deionized water. Sterile filter through 0.22 μm filter. Solution can be stored at room temperature.
8. Sorting medium: Hanks' balanced salt solution (HBSS) supplemented with 2 % FBS.
9. Transport solution: Hanks' balanced salt solution (HBSS) supplemented with 3.5 % BSA.
10. Digestion solution: 2 mg/mL Collagenase II dissolved in HBSS supplemented with 3.5 % BSA. Sterile filter through 0.45 μm filter before use (*see* **Note 1**).
11. Fisherbrand cell strainer 40, 70, and 100 μm.

2.2 FACS Staining

1. Compensation beads: OneComp eBeads.
2. Calcein blue stock solution: Prepare a 10 mM stock solution by diluting 1 mg calcein blue in 215 μL dimethylsulfoxid (DMSO).
3. Propidium iodide (PI). Use PI in a final concentration of 1 μg/mL.
4. Antibodies for FACS analysis of adipose tissue AdPCs: anti-mouse-Sca1 labeled with APC (clone D7; isotype Rat IgG2a) at a concentration of 0.5 μg/mL, anti-mouse-CD45 labeled with FITC (clone 30-F11; isotype Rat IgG2b) at a concentration of 2.5 μg/mL, anti-mouse-CD31 labeled with FITC (clone 390; isotype Rat IgG2a) at a concentration of 1 μg/mL.

2.3 In Vitro Differentiation of Murine Adipogenic Progenitor Cells

1. 24- or 48-well cell-culture plates.
2. Matrigel.
3. Epidermal growth factor (EGF).
4. Leukemia inhibitory factor (LIF).
5. Platelet-derived growth factor BB (PDGFR-BB).
6. Basic fibroblast growth factor (bFGF).
7. Gentamicin.
8. Growth medium for adipose tissue-derived AdPCs (modified from [37]): 60 % DMEM (low Glucose), 40 % MCDB201 Media, 2 % FBS, 100 U/mL Penicillin, 1000 U/mL Streptomycin, 1 nM Dexamethasone, 0.1 mM l-ascorbic acid-2P, 1× insulin-transferrin-selenium mix (ITS-MIX), 1× linoleic acid conjugated to bovine serum albumin (BSA). Ingredients are mixed, sterile filtered and the medium is stored at 4 °C.
9. Induction medium: Growth medium supplemented with: 5 μg/mL human insulin, 50 μM indomethacin, 1 μM dexamethasone, 0.5 μM isobutylmethylxanthine (IBMX), 1 nM T3.
10. Differentiation Medium: Growth medium supplemented with: 5 μg/mL human insulin, 1 nM T3.

3 Methods

3.1 Isolation of Adipogenic Progenitor Cells from Murine Adipose Tissues

1. Kill animals by cervical dislocation.
2. Spray dead animals with 70 % Ethanol. Dissect the interscapular brown adipose tissue (BAT) and the inguinal white adipose tissue pads (scWAT). For BAT, carefully remove residues of white adipose tissue and muscle from the brown adipose tissue with sterile scissors. Collect tissues into 50 mL conical tubes prefilled with a small volume of 3.5 % bovine serum albumin (BSA) in HBSS at 4 °C for short-term storage.
3. Mince the scWAT with sterile scissors for a few minutes until only small pieces are left. Note that mincing too extensively may lower the yield of AdPCs.
4. BAT is minced on a glass plate using a sharp razor blade until the sample is broken down to a pulp-like texture.
5. The minced tissues are collected in 50 mL conical tubes containing 3.5 % BSA in HBSS supplemented with 2 mg/mL Collagenase II and are sealed with parafilm. Digest the pieces for 1 h at 37 °C in a water bath under slight agitation. After the digestion process, the pieces should be mostly dissolved (*see* **Note 2**). To ensure a higher yield of AdPCs briefly but vigorously shake the conical tubes after 30 min of digestion.
6. Stop the digestion by adding 10 % FBS to the samples.
7. Centrifuge samples at 300 × *g* and at 4 °C for 10 min.
8. Aspirate the supernatant containing the mature adipocytes carefully. All adipocytes should be removed after this step. Some AdPCs may be dislodged from the pellet by the aspiration so the supernatant should not be removed completely.
9. Resuspend the pellet in 10 mL of Sorting medium (SM: 2 % FBS in HBSS) and filter through a 100 μm cell strainer into a fresh 50 mL conical tube. Wash the old tube with another 10 mL of SM and filter through the cell strainer.
10. Centrifuge samples at 300 × *g* and at 4 °C for 7 min.
11. Discard the supernatant and resuspend the pellet in 2 mL sterile ACK lysis buffer and incubate on ice for 3 min. Add 20 mL of SM to stop the lysis and filter the samples through a 40 μm cell strainer into a fresh 50 μL conical tube. Wash the old tube with 10 mL of SM and filter through the cell strainer.
12. Centrifuge samples at 300 × *g* and at 4 °C for 7 min and discard the supernatant. Resuspend the pellet in SM and transfer the cells into a sorting tube for staining.

3.2 Purification of Adipose-Tissue Progenitor Cells by FACS

AdPCs are purified using flow cytometry by staining with monoclonal, fluorochrome-tagged antibodies directed against specific cell surface markers. Antibodies are used against Sca1 as well as the hematopoietic and endothelial lineage (Lin) markers CD45 and CD31 (*see* **Note 3**).

1. Centrifuge the isolated cells at 300 × *g* for 5 min and resuspend the pellets in 150–250 μL of SM. Take small aliquots that are further used for control stainings (**step 2**).
2. Staining controls are used for voltage adjustments according to specific flow cytometer parameters. These may depend on the instrument used for sorting.
3. Compensation controls: For compensation prepare the following samples: One unstained sample, serving as a control for autofluorescence of the sorted cells, as well as control samples containing only one of the different fluorochromes used for sorting. Alternatively, beads may be used as compensation controls (except for unstained control), when the compensation controls using cells are not sufficient or feasible. In this regard, compensation beads that contain micro particles that bind to mouse isotype antibodies (positive compensation control) and micro particles with no binding capacity (negative compensation control) are recommended.
4. Fluorescence-minus-one (FMO) controls: FMOs are critical for setting the threshold of negative vs. positive signals. To determine the threshold of the signals and hence to differentiate between different cell populations, FMO controls are essential. These control samples contain all antibodies used for the staining scheme except the individual antibody for which the fluorescence threshold is to be determined.
5. For sample preparation add all antibodies at given concentrations (*see* Subheading 2.2) (*see* **Note 4**).
6. Samples and controls are incubated on ice and in the dark for at least 20 min.
7. Fill the sample tubes with SM and centrifuge the samples for 5 min at 300 × *g* and discard supernatant to remove unbound antibodies.
8. To avoid aggregates that might eventually clog the FACS instrument, filter samples through a 70 μm cell strainer right before sorting.
9. To distinguish between alive and dead cells add calcein blue (live cells) and propidium iodide (dead cells) to the samples. This may be done after the filtration step. The incubation time after the addition of PI and calcein should be kept constant for all samples.

10. For sorting cells with a FACS instrument start by compensating the spectral overlap of the fluorochromes (e.g., FITC and APC) using an unstained sample as control and the single color controls stained with only one fluorochrome. It should be noted that the compensation setting steps will differ significantly between different FACS instruments and should be performed by a trained expert.
11. After successful compensation, use the FMOs to distinguish between positive and negative staining signals for all the antibodies used in this protocol.
12. To identify and sort AdPCs, use a gating strategy as described in the following steps (also indicated in Fig. 1).
13. To exclude debris, select the population gated for the sideward scatter (SSC) and the forward scatter (FSC) (Fig. 1a, b). Note that debris locates in a cluster of SSC/FSC-negative particles just outside the R1-area in the lower left corner of the plot.
14. Cell duplets are excluded by selecting a distinct population gated for the forward scatter height (FSC-H) and the forward scatter area (FSC-A) (Fig. 1c, d).
15. Define the population of live cells by drawing a gate to include calcein blue positive cells. Exclude all cells that are positive for propidium iodide positive as this stains only dead cells (Fig. 1e, f). Note that a double-negative population of cellular debris usually occurs that should not be included in this gate.
16. Lastly, cells are sorted with regard to the staining signals for the cell surface markers, Sca1, CD45, and CD31. The FITC positive cells are either positive for CD31 or for CD45 and can thus be classified as endothelial cells or immune cells, respectively (Fig. 1g, h). The cells positive for the fluorophore APC express Sca1 (*see* **Note 5**). Additional fluorophores/antibodies may be included to more accurately define CD31- and CD45-expressing cells or additional subpopulations of cells depending on the available fluorescence channels of the flow cytometer. As discussed above, AdPCs are defined as the lineage, i.e., CD31 and CD45 (FITC) negative, and Sca1 (APC) positive population (*see* **Note 6**) (Fig. 1g, h). Different depots of adipose tissue contain distinct amounts of progenitor cells (Table 1).

3.3 Cell Culture and Adipogenic Differentiation of AdPCs (Fig. 2)

1. The cells are cultivated in 24- or 48-well plates.
2. For culture of AdPCs it is recommended to coat the cell culture dishes with Matrigel (*see* **Note 7**). Matrigel is diluted with cold DMEM (1 g/L glucose) to a final concentration of 2 % under sterile conditions. Matrigel is kept at 4 °C at all times, to avoid an uneven coating of the dishes. 2 % Matrigel solution is pipetted into the cell culture dish and aspirated again. The cell culture dishes are left to dry for several hours and should be exposed to UV-light for 20 min to ensure sterility.

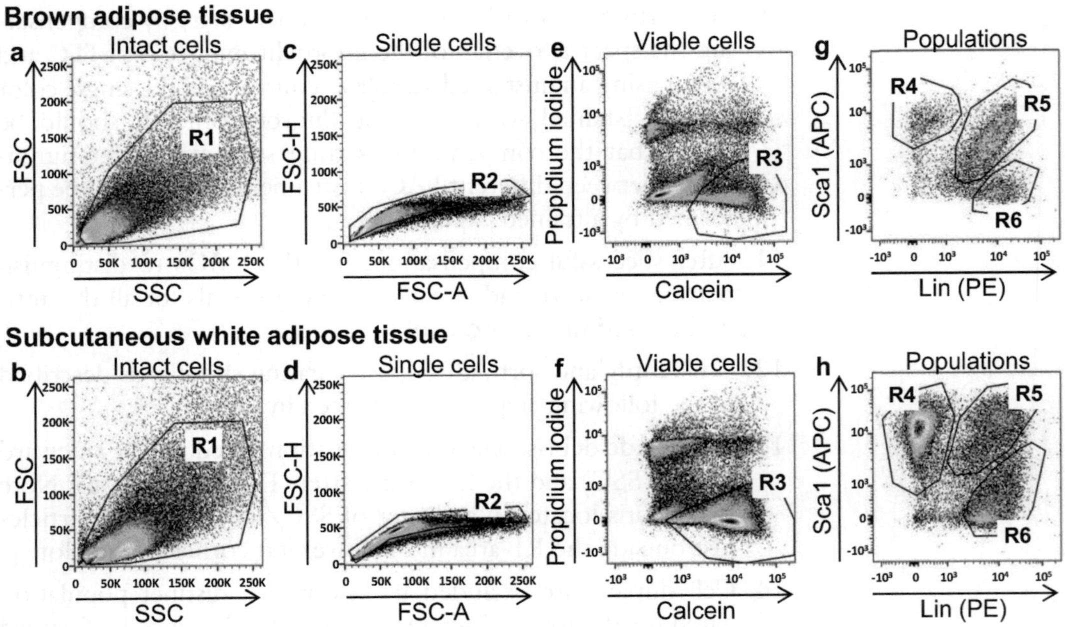

Fig. 1 Gating-strategy for the isolation of AdPCs from brown and subcutaneous white adipose tissue by FACS. (**a**, **b**) Draw a polygonal region (R1) around intact cells in the forward scatter (FSC) vs. side scatter (SSC) plot to exclude debris and retain intact cells. Note that a very large population of debris particles locates just outside the gate in the lower left corner. (**c**, **d**) Cell duplets are excluded by selecting a population (R2) identified by the forward scatter height (FSC-H) and the forward scatter area (FSC-A). (**e**, **f**) Calcein blue staining identifies alive cells, while staining with propidium iodide (PI) stains dead cells. Draw a gate as indicated around the Calcein$^+$/PI$^-$ population (R3). Note that two distinct populations in R3 of scWAT may be discernible, one of which contains (CD45$^+$) immune cells. (**g**, **h**) Adipogenic progenitors (AdPCs) are identified by selecting the Sca1$^+$/Lin$^-$cells (R4). CD31$^+$ endothelial cells are identified by selecting the Sca1$^+$/Lin$^+$ population (R5), whereas the majority of CD45$^+$ immune cells are Sca$^-$/Lin$^+$ (R6)

Table 1
Quantity of cell populations derived from adipose tissue after FACS analysis

	BAT	scWAT
Endothelial cells (CD31$^+$/Scal$^+$)	69.0 ± 1.7	13.6 ± 0.7
Hematopoietic cells (CD45$^+$/Scal$^-$)	8.6 ± 1.1	21.6 ± 2.4
Adipogenic progenitor cells (Scal$^+$/Lin$^-$)	15.5 ± 2.0	59.6 ± 2.3

Numbers depicted represent the percentage of cells normalized to viable cells ± SEM; N = 9–11

3. Purified AdPCs are centrifuged at 500 × g at 4 °C for 5 min.
4. Resuspend in Growth Medium supplemented with 10 ng/mL EGF, 10 ng/mL LIF, 10 ng/mL PDGFR-BB, and 5 ng/mL bFGF and seed 25,000 cells/well onto a Matrigel-covered

24-well cell culture plate. Supplement Growth medium with 50 μg/mL gentamicin to minimize the risk of bacterial contamination.

5. Expand the cells to 90–95 % confluency. Depending on the quality of the cell batch, this takes approximately one week.
6. Refresh the Growth medium supplemented with gentamicin every other day.
7. To perform an adipogenic differentiation assay, seed 15,000 cells/well onto a Matrigel-coated 48-well cell culture plate. The cells should then grow confluent in about three days which leaves time for potential pretreatments.
8. Upon confluence of the cells, induction of adipogenesis is achieved by adding the Induction medium and incubating the cells for 48 h (Fig. 2).
9. To further differentiate the cells, replace the Induction medium with Differentiation medium and refresh the medium every 48 h for one week (Fig. 2).
10. To enhance the differentiation efficacy of the cells, a treatment with the PPARγ agonist rosiglitazone can be performed. Supply the Differentiation medium with 1 μM of rosiglitazone starting at day 5 of the differentiation, i.e. after Induction medium is removed. Another possibility of enhancing the differentiation capacity is a pretreatment with 3.3 nM BMP7 for the first 72 h of differentiation during the preconfluent stages of the protocol. Both substances enhance brown adipogenesis and increase expression of the brown adipogenic marker UCP1 among other marker genes [19, 31] (Fig. 2).
11. Stimulation of the β3-adrenergic receptor (ADRB3) increases the expression of brown adipogenic markers and lipolytic activity. This can be achieved by exposing fully differentiated adipocytes to norepinephrine or the ADRB3-agonist, CL-316,243 prior to harvest [38]. Other possible stimulants include a short-term treatment with isoproterenol or membrane-permeable cAMP [31, 39]. To optimize the effects of the stimulation, add the substances 2–4 h prior to harvest of cells on the last day of differentiation [14, 40] (*see* **Note 8**) (Fig. 2).
12. For RNA isolation, aspirate culture medium carefully as differentiated cells dislodge easily and immediately add 0.2 mL/well of Trizol or similar reagent and lyse the cells by gently pipetting up and down. Alternatively, lipid accumulation could be assessed by applying a fixation agent followed by staining with Oil Red O [31].

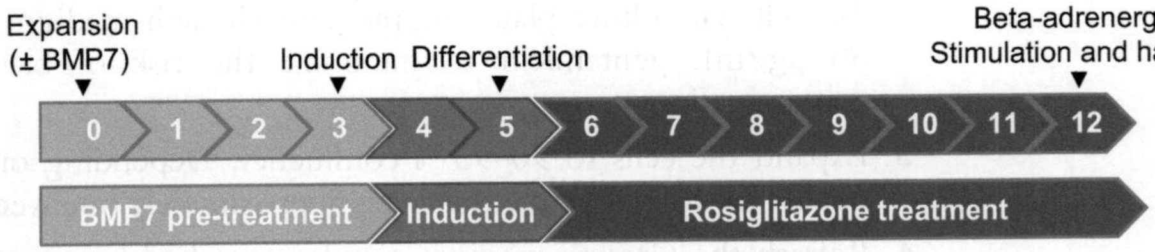

Fig. 2 Adipogenic differentiation of adipose tissue-derived AdPCs. The isolated and expanded cells are seeded into Matrigel-coated cell culture dishes. The BMP7 pretreatment is started after one day, giving the cells time to properly adhere. After 72 h the BMP7 containing medium is exchanged by adipogenic Induction medium. On the fifth day of differentiation, the cells are treated with Differentiation medium containing insulin and T3. Moreover, the Differentiation medium can be supplemented with the PPARγ agonist rosiglitazone for the remainder of the differentiation phase. The Differentiation medium is refreshed every other day and full adipogenic differentiation is achieved at day 12

4 Notes

1. The efficacy of the collagenase batch should be determined prior to use. Differences in enzyme activity may affect the dissociation of the tissue or the viability of the cells, both resulting in a lower yield of AdPCs.
2. The digestion time may need to be adjusted depending on the collagenase batch used and should be determined empirically.
3. Other cell populations rather than AdPCs may be identified and isolated by staining additional surface markers. For instance, the CD45 positive immune cells may be further subdivided by adding the macrophage marker EGF-like module-containing mucin-like hormone receptor-like 1 (F4/80) or the T-cell markers CD4 and CD8 and adapting the gating strategy [41].
4. To optimize the yield and minimize the occurrence of false-positive signals, the concentration of all antibodies used should be optimized by titration.
5. scWAT contains a higher percentage of AdPCs in comparison to BAT, leading to a higher yield in scWAT (Table 1).
6. The differentiation protocol is optimized for brown adipogenesis and thus the differentiation rate of AdPCs derived from scWAT is lower compared to AdPCs isolated from BAT. Supplementation with the enhancing factors is recommended to compensate this effect.
7. Keep Matrigel on ice at any time to avoid the solidification of the pure Matrigel. The diluted Matrigel should be kept on ice and can be stored for 2 weeks at 4 °C.
8. Norepinephrine is very unstable. Solutions should be kept frozen and only used for a limited amount of time. Aliquot stock solution before freezing and use each aliquot only once. Do not thaw and re-freeze for later use. Norepinephrine should not be exposed to light.

Acknowledgments

This work was supported by grants from the German Research Foundation (DFG; grant # SCHU 2445/2-1) and the European Research Council (grant # ERC-StG-2012-311082) to T.J.S. The authors gratefully acknowledge support from the German Center for Diabetes Research (DZD).

References

1. World Health Organization (2014) Obesity and overweight fact sheet N°311.http://www.who.int/mediacentre/factsheets/fs311/en/
2. Bluher M (2014) Adipokines—removing road blocks to obesity and diabetes therapy. Mol Metab 3:230–240
3. Kershaw EE, Flier JS (2004) Adipose tissue as an endocrine organ. J Clin Endocrinol Metab 89:2548–2556
4. Zhang Y, Proenca R, Maffei M et al (1994) Positional cloning of the mouse obese gene and its human homologue. Nature 372:425–432
5. Nedergaard J, Bengtsson T, Cannon B (2007) Unexpected evidence for active brown adipose tissue in adult humans. Am J Physiol Endocrinol Metab 293:E444–E452
6. Cypess AM, Lehman S, Williams G et al (2009) Identification and importance of brown adipose tissue in adult humans. N Engl J Med 360:1509–1517
7. Saito M, Okamatsu-Ogura Y, Matsushita M et al (2009) High incidence of metabolically active brown adipose tissue in healthy adult humans: effects of cold exposure and adiposity. Diabetes 58:1526–1531
8. van Marken Lichtenbelt WD, Vanhommerig JW, Smulders NM et al (2009) Cold-activated brown adipose tissue in healthy men. N Engl J Med 360:1500–1508
9. Virtanen KA, Lidell ME, Orava J et al (2009) Functional brown adipose tissue in healthy adults. N Engl J Med 360:1518–1525
10. Zingaretti MC, Crosta F, Vitali A et al (2009) The presence of UCP1 demonstrates that metabolically active adipose tissue in the neck of adult humans truly represents brown adipose tissue. FASEB J 23:3113–3120
11. Yoneshiro T, Aita S, Matsushita M et al (2013) Recruited brown adipose tissue as an antiobesity agent in humans. J Clin Invest 123:3404–3408
12. Matsushita M, Yoneshiro T, Aita S et al (2013) Impact of brown adipose tissue on body fatness and glucose metabolism in healthy humans. Int J Obes (Lond) 38:812–817
13. Bakker LEH, Boon MR, van der Linden RAD et al (2014) Brown adipose tissue volume in healthy lean south Asian adults compared with white Caucasians: a prospective, case-controlled observational study. Lancet Diabetes Endocrinol 2:210–217
14. Cannon B, Nedergaard J (2004) Brown adipose tissue: function and physiological significance. Physiol Rev 84:277–359
15. Lichtenbelt WV, Kingma B, van der Lans A et al (2014) Cold exposure—an approach to increasing energy expenditure in humans. Trends Endocrinol Metab 25:165–167
16. Schulz TJ, Tseng YH (2013) Brown adipose tissue: development, metabolism and beyond. Biochem J 453:167–178
17. Rosen ED, Spiegelman BM (2014) What we talk about when we talk about fat. Cell 156:20–44
18. Ishibashi J, Seale P (2010) Medicine. Beige can be slimming. Science 328:1113–1114
19. Petrovic N, Walden TB, Shabalina IG et al (2010) Chronic peroxisome proliferator-activated receptor gamma (PPAR gamma) activation of epididymally derived white adipocyte cultures reveals a population of thermogenically competent, UCP1-containing adipocytes molecularly distinct from classic brown adipocytes. J Biol Chem 285:7153–7164
20. Shabalina IG, Petrovic N, de Jong JM et al (2013) UCP1 in brite/beige adipose tissue mitochondria is functionally thermogenic. Cell Rep 5:1196–1203
21. Kajimura S, Seale P, Spiegelman BM (2010) Transcriptional control of brown fat development. Cell Metab 11:257–262
22. Kajimura S, Seale P, Kubota K et al (2009) Initiation of myoblast to brown fat switch by a PRDM16-C/EBP-beta transcriptional complex. Nature 460:1154–1158
23. Seale P, Bjork B, Yang W et al (2008) PRDM16 controls a brown fat/skeletal muscle switch. Nature 454:961–967

24. Seale P, Conroe HM, Estall J et al (2011) Prdm16 determines the thermogenic program of subcutaneous white adipose tissue in mice. J Clin Invest 121:96–105
25. Vitali A, Murano I, Zingaretti MC et al (2012) The adipose organ of obesity-prone C57BL/6J mice is composed of mixed white and brown adipocytes. J Lipid Res 53:619–629
26. Himms-Hagen J, Melnyk A, Zingaretti MC et al (2000) Multilocular fat cells in WAT of CL-316243-treated rats derive directly from white adipocytes. Am J Physiol Cell Physiol 279:C670–C681
27. Barbatelli G, Murano I, Madsen L et al (2010) The emergence of cold-induced brown adipocytes in mouse white fat depots is determined predominantly by white to brown adipocyte transdifferentiation. Am J Physiol Endocrinol Metab 298:E1244–E1253
28. Schulz TJ, Huang P, Huang TL et al (2013) Brown-fat paucity due to impaired BMP signalling induces compensatory browning of white fat. Nature 495:379–383
29. Weisberg SP, McCann D, Desai M et al (2003) Obesity is associated with macrophage accumulation in adipose tissue. J Clin Invest 112:1796–1808
30. Rodeheffer MS, Birsoy K, Friedman JM (2008) Identification of white adipocyte progenitor cells in vivo. Cell 135:240–249
31. Schulz TJ, Huang TL, Tran TT et al (2011) Identification of inducible brown adipocyte progenitors residing in skeletal muscle and white fat. Proc Natl Acad Sci USA 108: 143–148
32. Tang W, Zeve D, Suh JM et al (2008) White fat progenitor cells reside in the adipose vasculature. Science 322:583–586
33. Berry DC, Stenesen D, Zeve D et al (2013) The developmental origins of adipose tissue. Development 140:3939–3949
34. Berry R, Rodeheffer MS (2013) Characterization of the adipocyte cellular lineage in vivo. Nat Cell Biol 15:302–308
35. Tseng YH, Kokkotou E, Schulz TJ et al (2008) New role of bone morphogenetic protein 7 in brown adipogenesis and energy expenditure. Nature 454:1000–1004
36. Vegiopoulos A, Muller-Decker K, Strzoda D et al (2010) Cyclooxygenase-2 controls energy homeostasis in mice by de novo recruitment of brown adipocytes. Science 328:1158–1161
37. Steenhuis P, Pettway GJ, Ignelzi MA Jr (2008) Cell surface expression of stem cell antigen-1 (Sca-1) distinguishes osteo-, chondro-, and adipoprogenitors in fetal mouse calvaria. Calcif Tissue Int 82:44–56
38. Himms-Hagen J, Cui J, Danforth E Jr et al (1994) Effect of CL-316,243, a thermogenic beta 3-agonist, on energy balance and brown and white adipose tissues in rats. Am J Physiol 266:R1371–R1382
39. Klaus S, Choy L, Champigny O et al (1994) Characterization of the novel brown adipocyte cell line HIB 1B. Adrenergic pathways involved in regulation of uncoupling protein gene expression. J Cell Sci 107:313–319
40. Lehr L, Canola K, Asensio C et al (2006) The control of UCP1 is dissociated from that of PGC-1alpha or of mitochondriogenesis as revealed by a study using beta-less mouse brown adipocytes in culture. FEBS Lett 580:4661–4666
41. Orr JS, Kennedy AJ, Hasty AH (2013) Isolation of adipose tissue immune cells. J Vis Exp 22(75):e50707

Chapter 5

Immuno-Magnetic Isolation and Thermogenic Differentiation of White Adipose Tissue Progenitor Cells

Rohollah Babaei, Irem Bayindir-Buchhalter, Irina Meln, and Alexandros Vegiopoulos

Abstract

Appropriate cell models are necessary for the investigation of thermogenic beige adipocyte differentiation from progenitor cells. Here, we describe a primary cell culture method that is based on defined progenitor cells from murine white adipose tissue and aims at minimizing confounding factors including cell heterogeneity and nonphysiological differentiation inducers. Adipocyte progenitor cells are enriched by immunomagnetic separation, expanded minimally, and induced for beige adipocyte differentiation with carbaprostacyclin, a stable analogue of the endogenous mediator PGI_2.

Key words Adipocyte progenitors, Thermogenic adipocytes, Beige adipocytes, Browning, Adipogenic differentiation, Primary cell isolation, Carbaprostacyclin, cPGI2

1 Introduction

Thermogenic adipocytes, i.e., brown and beige adipocytes, are able to influence systemic energy balance and metabolic homeostasis [1]. Through the controlled generation of heat they contribute to thermoregulation and adaptation to cold environments. Moreover, increased abundance of thermogenic adipocytes confers protection from obesity, dyslipidemia, and type-2 diabetes, mediated by metabolic and endocrine pathways [2, 3]. For this reason, approaches for the enhancement of thermogenic capacity are being actively pursued toward the development of new therapeutic interventions for metabolic disease. Detailed understanding of the processes that control the abundance of thermogenic adipocytes is crucial in this direction.

In rodents, prolonged cold exposure and other environmental conditions lead to increased abundance of beige adipocytes in white fat depots [1, 3]. To a substantial extent, these cells originate from immature progenitor cells, which respond to systemic stress

Jun Wu (ed.), *Thermogenic Fat: Methods and Protocols*, Methods in Molecular Biology, vol. 1566,
DOI 10.1007/978-1-4939-6820-6_5, © Springer Science+Business Media LLC 2017

signals to differentiate toward an oxidative/thermogenic phenotype [1, 2]. Adult human adipose tissue progenitor cells have been shown to have the potential to form thermogenic adipocytes [1, 3]. Therefore, the induction of thermogenic differentiation in progenitor cells represents an attractive therapeutic target. In recent years, there has been tremendous progress in the identification of regulatory nodes controlling differentiation and function of thermogenic adipocytes [2]. However, with a few exceptions, the signaling pathways regulating progenitor-mediated beige adipocyte formation downstream of physiological mediators remain to be determined.

Besides the lack of genetic tools for specifically manipulating adipocyte progenitor cells in mice, the paucity of physiological cell models has been an obstacle for the investigation of progenitor regulation. Preadipocyte or mesenchymal cell lines have contributed substantially to the discovery of master regulatory factors of adipogenic/thermogenic differentiation. However, in many cases, their properties do not reflect their in vivo counterparts, in particular with regards to cell cycle regulation and responsiveness to physiological stimuli. On the other hand, most studies on primary cells have relied on culturing the rather heterogeneous stromal-vascular fraction (SVF) of adipose tissue, with undefined and varying content of non-progenitor cells, including endothelial or non-adipogenic fibroblasts, which can confound the findings and increase variability. Nevertheless, the key determinant of cell culture models for beige adipocyte differentiation is likely to be the specific stimulus of the thermogenic phenotype. In this respect, treatment of progenitor cells from white fat with Bone morphogenetic protein 7 (Bmp7), which promotes commitment to the thermogenic program, represents a valid model, since the Bmp7 signaling pathway has been shown to be of central importance in vivo [4]. The most commonly used stimulus is the activation of the PPARγ nuclear receptor by synthetic agonists (thiazolidinediones (TZDs)), mostly rosiglitazone [5]. This is consistent with the central role of PPARγ in the regulation of thermogenic gene expression. However, TZDs trigger supra-physiological activation of PPARγ and potently promote general adipogenesis in parallel to beige differentiation [6, 7].

With the aim of better approximation of physiological conditions, we have recently developed a cell model based on defined primary progenitor cells and the inducer carbaprostacyclin (cPGI$_2$) [8]. We have chosen murine adipose tissue as a cell source as opposed to human, due to the broader accessibility of fresh biopsies, the availability of genetically modified strains, and the reduced variability. To further reduce variability, we have employed purified collagenase/proteinase enzyme preparations with defined activities for the generation of single cell suspensions from adipose tissue biopsies. The method is based on a purified cell population defined by the surface marker combination Lin(Ter119/Cd45/CD31)$^-$CD29$^+$CD34$^+$Sca1$^+$, which was shown by Rodeheffer *et al.* to contain all adipogenic cells [9]. Therefore, it is likely that this population includes the committed beige precursors described by Wu et al. [10]. Importantly, this population contains most of the Pdgfra$^+$ and Pdgfrb$^+$ cells, which cover a majority of the thermogenic adipocyte precursors in white fat, as proven by lineage tracing [8, 11, 12]. The protocol presented here employs a simplified immuno-magnetic purification of Lin$^-$Sca1$^+$ cells, which we have shown to have beige adipocyte potential comparable to the above FACS-sorted population [8]. This involves the removal of Ter119$^+$ erythrocytes, CD45$^+$ leucocytes, and CD31$^+$ endothelial cells by antibody labeling and magnetic negative selection (Fig. 1). In a second step, Sca1$^+$ cells are purified by positive selection. Following a short growth phase in culture without passaging, the cells are induced to differentiate by basic adipogenic factors and cPGI$_2$, a stable analogue of prostacyclin (PGI$_2$) (Fig. 2). In vivo, the rate-limiting reaction in the biosynthesis of PGI$_2$ is catalyzed by the

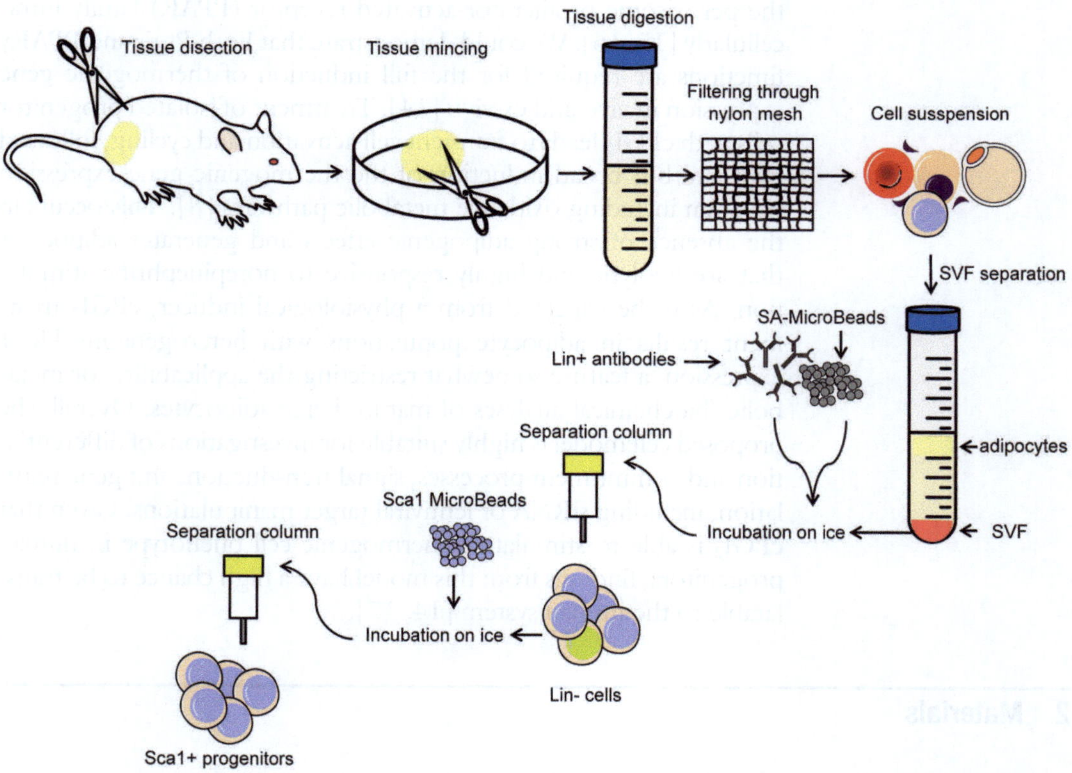

Fig. 1 Schematic of the progenitor isolation procedure from mouse fat tissue. Mouse posterior subcutaneous fat pads are dissected and minced manually with scissors. The tissue is digested in collagenase solution. Digested tissue is strained through nylon mesh and the stromal-vascular fraction (SVF) is separated from mature adipocytes by centrifugation. Lin$^+$ cells are removed from the SVF fraction by biotinylated anti-Ter119/CD31/CD45 antibody labeling and magnetic depletion using SA-MicroBeads. Sca1$^+$ progenitors are purified from the Lin$^-$ population using Sca1-MicroBeads

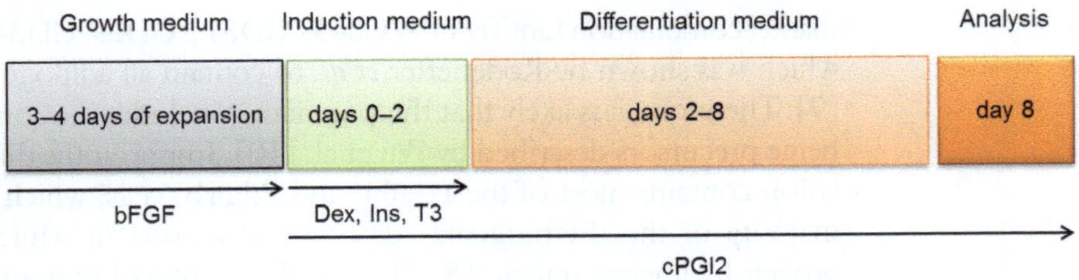

Fig. 2 Schematic of the culture and differentiation procedure for freshly isolated adipose tissue progenitors. Isolated Lin$^-$Sca1$^+$ progenitors are seeded on laminin-coated 24-well plates and maintained in growth medium including mbFGF, typically for 3–4 days. Beige adipogenic differentiation is induced by induction medium in the presence of cPGI$_2$ for 2 days. Full differentiation is achieved typically within additional 6 days in differentiation medium including cPGI$_2$

cyclooxygenases 1 and 2 (COX-1 and -2). We and others have previously revealed the essential role of COX-2 in the thermogenic remodeling of white adipose tissue downstream of β-adrenergic signaling [13, 14]. PGI$_2$ signaling is mediated through binding and activation of the G-protein coupled IP receptor (Ptgir) as well as members of the peroxisome proliferator-activated receptor (PPAR) family intracellularly [15, 16]. We could demonstrate that both Ptgir and PPARγ functions are required for the full induction of thermogenic gene expression in vivo and ex vivo [14]. Treatment of isolated progenitor cells with cPGI$_2$ leads to transient cell activation and cycling, followed by a late but broad induction of the thermogenic gene expression program including oxidative metabolic pathways [8]. This occurs in the absence of strong adipogenic effects and generates adipocytes that are lipolytic and highly responsive to norepinephrine stimulation. As to be expected from a physiological inducer, cPGI$_2$ treatment results in adipocyte populations with heterogeneous Ucp1 expression, a feature somewhat restricting the applicability for metabolic/biochemical analyses of mature beige adipocytes. Overall, the proposed cell model is highly suitable for investigations of differentiation and commitment processes, signal transduction, and gene regulation, including siRNA or lentiviral target manipulations. Given that cPGI$_2$ is able to stimulate a thermogenic cell phenotype in human progenitors, findings from this model have a high chance to be translatable to the human system [14, 17].

2 Materials

2.1 Tissue Digestion and Lin$^-$Sca1$^+$ Progenitor Cell Isolation

1. 10 cm petri dishes.
2. 6-well plates.
3. 37 °C water bath.
4. 37 °C shaker.
5. Swing-rotor centrifuge.
6. Ethanol 70 %.

7. Bench liner.
8. 0.22 μ filter.
9. 300 μ nylon mesh.
10. Streptavidin MicroBeads (SA-MicroBeads, Miltenyi Biotec).
11. Anti-Sca-1 (non-HSC) MicroBeads, mouse (Miltenyi Biotec, 130-106-641).
12. MS column (Miltenyi Biotec).
13. Pre-Separation Filters 30 μ(Miltenyi Biotec).
14. OctoMACS Separator Starter Kit (Miltenyi Biotec).
15. Anti-CD16/32 (FcBlock), (clone 93).
16. Anti-Ter119-biotin (clone TER-119).
17. Anti-CD31-biotin (clone 390).
18. Anti-CD45-biotin (clone 30-F11).
19. Bovine serum albumin (BSA) buffer: Dulbecco's phosphate-buffered saline (PBS, without Ca, Mg), BSA 0.5 %, Ethylenediaminetetraacetic acid (EDTA) 1 mM (*see* **Note 1**).
20. Digestion solution: Hanks buffer (KCl 0.4 g/L, KH_2PO_4 0.06 g/L, $NaHCO_3$ 0.35 g/L, NaCl 8 g/L, Na_2HPO_4 0.048 g/L, D-Glucose 1 g/L, endotoxin free), DNase I (from bovine pancreas, >2000 U/mg) 50 μg/mL, $CaCl_2$4 mM, Collagenase (purified CLSPA, Worthington) 133 U/mL (0.1 W.u./mL) (*see* **Note 2**), Dispase (NPRO, Worthington) (*see* **Note 3**) 2.4 U/mL (*see* **Note 4**).

2.2 Cell Culture and Adipogenic Differentiation

1. 24-well laminin-coated plate (*see* **Note 5**).
2. Growth medium: DMEM high glucose, Fetal Bovine Serum (FBS) (*see* **Note 6**) 10 %, Penicillin/Streptomycin (Pen.Strep) 1 %, Murine rec. basic fibroblast growth factor (mbFGF) 10 ng/mL.
3. Induction medium: DMEM, FBS 10 %, Pen.Strep 1 %, Dexamethasone (Dex) 500 nM, human recombinant (yeast) insulin (Ins) 1 μg/mL, Triiodothyronine (T3) 3 nM, Carbaprostacyclin ($cPGI_2$) 1 μM (*see* **Note 7**).
4. Differentiation medium: DMEM, FBS 5 %, Pen.Strep, Ins 1 μg/mL, T3 3 nM, $cPGI_2$ 1 μM.

3 Methods

3.1 Tissue Dissection

1. Prepare all buffers, solutions and instruments before dissecting the tissue (*see* **Note 8**).
2. Sacrifice the mouse by cervical dislocation (*see* **Note 9**). Wet the fur with 70 % ethanol and place the mouse on a clean bench liner.

3. Make an incision at the skin on the lower back of mouse. Insert the scissors under the skin and open. Close and remove the scissors. Repeat this several times in different directions until the skin is detached from the body. Dissect a skin flap from the lower back and fold it onto the upper back.
4. Dissect the right and left posterior subcutaneous fat pads (gluteal + inguinal fat) and keep in PBS on ice.

3.2 Tissue Digestion

1. Prepare the digestion solution. Keep the solution on ice until needed. Transfer the fat pads in a 10 cm-petri dish and mince with very sharp scissors until fat pieces are not easily visible in the homogenate. The inguinal lymph nodes should be removed during this process.
2. Aliquot the digestion solution into 15 mL-Falcon tubes (*see* **Note 10**). Transfer the minced fat to Falcon tubes containing digestion solution, and mix by inverting the tube. Incubate the mixture in a pre-warmed water bath at 37 °C for 10 min. Complete the digestion by incubating the mixture in a pre-warmed shaker at 37 °C for 1 h at 80 rpm (*see* **Note 11**).

3.3 Lin^-Sca1^+ Progenitor Cell Isolation

1. Remove the nondigested particles by straining through a 300 µ-nylon mesh, and collect the flow-through in a 50 mL-falcon tube (*see* **Note 12**).
2. Transfer the suspension to a 15 mL-Falcon tube (two mice per tube) and spin down on a swing rotor at 145 × *g* for 10 min at room temperature (RT).
3. Mature adipocytes should be floating as a top layer, whereas the SVF is pelleted at the bottom of the tube. The mature adipocyte fraction can be collected for independent experiments using a 1 mL micropipette, or it should be aspirated/discarded at this step.
4. Aspirate the supernatant using a glass Pasteur pipette connected to a vacuum pump (*see* **Note 13**).
5. Resuspend the pellet very gently in 1 mL BSA buffer using a 1 mL-micropipette. Add 11 mL BSA buffer and mix by inverting the tube. Spin down on a swing rotor at 300 × *g* for 5 min at RT (*see* **Note 14**).
6. Aspirate the supernatant and resuspend the pellet in 1 mL BSA buffer.
7. Add 20 µL of anti-CD16/32, mix by taping at the bottom of the tube, and incubate on ice for 10 min.
8. Cool down the centrifuge to 4 °C.
9. Add 10 µL of anti-Ter119-bio, 10 µL of anti-CD31-bio, and 10 µL of anti-CD45-bio antibodies, mix by taping at the bottom of the tube, and incubate for 30 min on ice (*see* **Note 15**).

10. Spin down at 300 × *g* for 5 min at 4 °C. Aspirate the supernatant, resuspend the pellet in 1 mL BSA buffer, and mix gently with a 1 mL-micropipette. Add 9 mL BSA buffer and mix by inverting the tube.
11. Spin down at 300 × *g* for 5 min at 4 °C. Aspirate the supernatant and resuspend the pellet in BSA buffer (90 μL/mouse).
12. Add 10 μL/mouse SA-MicroBeads and mix by taping at the bottom of the tube. Incubate for 15 min at 4–8°C.
13. Meanwhile, assemble the OctoMACS separator, and mount the MS column incl. a pre-separation filter on it (*see* **Note 16**).
14. Add 1 mL BSA buffer to each column. Collect the flow-through in a petri dish and discard.
15. Add 4 mL BSA buffer to the samples and mix by inverting the tube. Spin down at 300 × *g* for 5 min at 4 °C.
16. Aspirate the supernatant and resuspend the pellet in 1 mL BSA buffer per column (for samples obtained from two mice you will need one column).
17. Provide 15 mL-Falcon tubes under the columns to collect the Lin^- flow-through.
18. Apply 1 mL of the cell suspension to each column and collect the flow-through.
19. Wash the columns three times with 500 μL BSA buffer and collect the flow-through from each wash.
20. Spin down the collected flow-through at 300 × *g* for 5 min at 4 °C.
21. Aspirate the supernatant, and resuspend the cells in BSA buffer (20 μL per mouse, i.e., for samples obtained from four mice you will need 80 μL BSA buffer).
22. Add 5 μL/mouse of anti-Sca-1 MicroBeads and mix by taping at the bottom of the tube. Incubate at 4–8°C for 15 min (*see* **Note 17**).
23. Add 1 mL of BSA buffer to the cells and spin down at 300 × *g* for 5 min at 4 °C.
24. Meanwhile, mount new columns and pre-separation filters on OctoMACS. Prepare the columns by applying 500 μL of BSA buffer. Collect the flow- through in a petri dish and discard.
25. Aspirate the cell supernatant and resuspend the cells in 500 μL BSA buffer per column using a 1 mL micropipette. Cells obtained from four mice will be applied to one column.
26. Apply 500 μL of cell suspension to each column and collect the flow-through in a petri dish to discard ($Sca1^-$ fraction).
27. Wash the column three times with 500 μL BSA buffer (discard flow-through).

28. When the column is fully drained after the washing steps, remove the pre-separation filter from the column, unmount the column from OctoMACS, and mount it on a 15 mL Falcon tube.
29. Add 1 mL of BSA buffer to column, insert the plastic piston supplied with the column into the column, press to elute the cells from column, and collect the flow-through in the tube (without applying excessive pressure).
30. Spin the cells down at 300 × *g* for 5 min at RT. Aspirate the buffer and resuspend the cells in a 1 mL pre-warmed growth medium.
31. Mix 40 μL of cells with 10 μL trypan blue, and count the cells using a Neubauer hemocytometer (*see* **Note 18**).

3.4 Cell Culture and Differentiation

1. Allow the laminin-coated plates to reach RT.
2. Prepare a suspension of freshly isolated $Lin^{-}Sca1^{+}$ cells in growth medium for seeding at 0.9×10^4 cells/cm^2 (1.8×10^4 cells in 500 μL/well for 24-wells) (*see* **Notes 19** and **20**).
3. Transfer appropriate volume of cell suspension to the center of each well, using a 1 mL micropipette (*see* **Note 21**).
4. Incubate the plate in an incubator at 37 °C, supplemented by 5 % CO_2, for 2 days.
5. Look under the microscope and confirm that the cells are settled well. Replace the growth medium by fresh and incubate for 1 day. The time point of differentiation induction (Day 0) is determined based on cell density (*see* **Note 22**).
6. Day 0: Replace the growth medium by induction medium and incubate for 24 h.
7. Day 1: Replace the induction medium by fresh and incubate for 24 h.
8. Day 2: Replace the induction medium by differentiation medium.
9. Day 3–7: Replace the medium by fresh differentiation medium daily (*see* **Note 23**).
10. Day 8: Process the cells according to the analytical method of choice (*see* **Note 24**). For example, Ucp1 protein can be detected by immunofluorescence (Fig. 3) (*see* **Notes 25** and **26**).

4 Notes

1. Prepare 50 mL of BSA buffer, before dissecting the tissue from mice. This volume of buffer is enough for digesting the fat pad obtained from four mice.

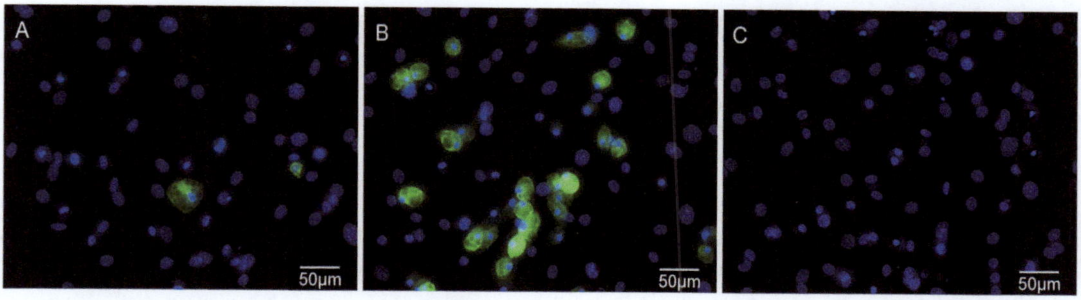

Fig. 3 Ucp1 immunofluorescence images at 8 days of differentiation. Lin$^-$Sca1$^+$ progenitors were cultured and differentiated in the absence (**a**) or presence (**b, c**) of cPGI$_2$. At day 8 of differentiation the cells were fixed and stained using anti-Ucp1 antibody (*green*) (**a, b**) and DAPI (*blue*). The Ucp1 signal was detected by Alexa Fluor-conjugated secondary antibody. (**c**) The primary anti-Ucp1 antibody was omitted from the protocol

2. Prepare 2.333 U/mL (corresponding to approx. 1.75 Wunsch units (W.u.)/mL) stock solution in Hanks buffer, aliquot in 1-mL tubes, and store at −20 °C.
3. Prepare 47.8 U/mL stock solution in Hanks buffer, aliquot in 1-mL tube, and store at −20 °C.
4. Thaw on ice the reagents needed for digestion solution, before dissecting the tissue from mice. The digestion solution should be prepared freshly, just after dissecting the tissues, and before mincing. Prepare 4 mL of digestion solution per depot/mouse (but digest maximally 1 g of tissue in 4 mL digestion solution). High tissue density in the digestion solution can reduce cell recovery.
5. As an alternative to commercially available laminin-coated plates (BD BioCoat™), standard tissue culture plates can be coated in the lab. To coat the wells, prepare 4 μg/cm^2 laminin (8 μg/well of 24-well plate) in DMEM (250 μL/well). Cover the wells by laminin and incubate at RT for 1 h. Aspirate and wash gently with 500 μL PBS.
6. The origin and batch of FBS has substantial influence on the efficiency of adipogenic differentiation incl. beige adipogenic capacity. FBS batches should be tested for comparable adipogenic capacity. Low-endotoxin FBS should be preferred.
7. Omitting cPGI$_2$ from the media results in the formation of adipocytes resembling mainly white adipocytes (based on thermogenic marker expression). Addition of equivalent amount of ethanol in place of cPGI$_2$ is required to represent the vehicle controls for cPGI$_2$-treated cultures.
8. Warm up the water bath and the shaker to 37 °C before dissecting the tissue from mice.
9. The protocol is well applicable for C57BL/6 and NMRI mice. It is noteworthy though that NMRI can be advantageous due to higher yield of progenitor cells and higher capacity for white as well as beige adipocyte differentiation.

10. Don't use more than 8 mL per 15 mL tube, to leave enough space for shaking.
11. Set the tube in the shaker in a horizontal angle and check the digestion every 10 min. Don't shake the mixture more than 1 h. Longer shaking durations and higher shaking intensities may lead to clotting in the later steps of cell isolation, reduced cell recovery and viability. Therefore, the shaking conditions and duration should be tested/optimized in each lab.
12. To avoid contamination, cut the nylon mesh in 8 × 8 cm pieces and autoclave in water.
13. Keep the glass pipette at the surface of the supernatant, since the solution could be viscous leading to aspiration and loss of SVF cells from the pellet.
14. All resuspensions using 1 mL micropipette should be done very gently to avoid the clotting of the cells.
15. Mix the cell suspension every 10 min by taping at the bottom of the tube.
16. In this step, use one column for material obtained from two mice.
17. In this step, one column should be used for samples obtained from four mice. Therefore, for each column, the cells should be resuspended in 80 μL of BSA buffer and mixed with 20 μL of anti-Sca-1 MicroBeads.
18. The cell yield varies strongly with mouse strain, sex, and age. Typical yields of $Lin^{-}Sca1^{+}$ cells per female mouse for the C57BL/6 and NMRI strains are $1.5–3 \times 10^5$ and $3–6 \times 10^5$ cells, respectively.
19. Passaging of $Lin^{-}Sca1^{+}$ cells results in marked decrease in the differentiation capacity.
20. Per 24-well plate prepare an excess of three wells.
21. If the well is not covered by medium, move the plate very gently to cover the well. Avoid aggressive 8-shape or cross-shape movement, since it might lead to the aggregation of the cells in the middle of the well.
22. The cell density at the time point of induction has strong influence on the differentiation efficiency. The induction of differentiation should be applied when the cells have reached the confluence of 70–80 %. If the cells are not confluent enough, incubate them overnight without replacing the medium. However, overgrowing of the cultures may lead to complete loss of differentiation capacity.
23. To avoid detachment and rupture of maturing adipocytes, keep the cell layer covered with medium at all times when replacing media (i.e., approx. 50 μL of medium should remain in the well after aspiration).

24. The cells can be cultured for additional days but depending on the conditions, some deterioration of cell viability and lipid content may begin to occur.
25. Depending on factors including mouse strain, FBS properties, and cell density, FACS-purified $Lin^{-}CD29^{+}CD34^{+}Sca1^{+}$ cells may yield higher adipogenic capacity compared to MACS-purified $Lin^{-}Sca1^{+}$ cells. However, the inducibility of beige adipocyte differentiation by $cPGI_2$ does not seem to be affected.
26. Differentiation efficiency including thermogenic differentiation (Ucp1) can be moderate compared to IBMX/rosiglitazone-treated cells, depending on factors including mouse strain, age, FBS properties, and cell density. However, $cPGI_2$ treatment is not compatible with IBMX or rosiglitazone stimulation.

Acknowledgments

This work was supported by the Deutsche Forschungsgemeinschaft (HE 3260/8-1), the Helmholtz Association ("Metabolic Dysfunction"), and the Human Frontier Science Program (RGY0082/2014).

References

1. Diaz MB, Herzig S, Vegiopoulos A (2014) Thermogenic adipocytes: from cells to physiology and medicine. Metabolism 63(10):1238–1249. doi:10.1016/j.metabol.2014.07.002
2. Kajimura S, Spiegelman BM, Seale P (2015) Brown and Beige fat: physiological roles beyond heat generation. Cell Metab 22(4):546–559. doi:10.1016/j.cmet.2015.09.007
3. Sidossis L, Kajimura S (2015) Brown and beige fat in humans: thermogenic adipocytes that control energy and glucose homeostasis. J Clin Invest 125(2):478–486. doi:10.1172/JCI78362
4. Schulz TJ, Huang TL, Tran TT, Zhang H, Townsend KL, Shadrach JL, Cerletti M, McDougall LE, Giorgadze N, Tchkonia T, Schrier D, Falb D, Kirkland JL, Wagers AJ, Tseng YH (2011) Identification of inducible brown adipocyte progenitors residing in skeletal muscle and white fat. Proc Natl Acad Sci USA 108(1):143–148. doi:10.1073/pnas.1010929108
5. Petrovic N, Walden TB, Shabalina IG, Timmons JA, Cannon B, Nedergaard J (2010) Chronic peroxisome proliferator-activated receptor gamma (PPARgamma) activation of epididymally derived white adipocyte cultures reveals a population of thermogenically competent, UCP1-containing adipocytes molecularly distinct from classic brown adipocytes. J Biol Chem 285(10):7153–7164. doi:10.1074/jbc.M109.053942
6. Lehmann JM, Moore LB, Smitholiver TA, Wilkison WO, Willson TM, Kliewer SA (1995) An antidiabetic thiazolidinedione is a high-affinity ligand for peroxisome proliferator-activated receptor gamma(Ppar-Gamma). J Biol Chem 270(22):12953–12956
7. Samarasinghe SP, Sutanto MM, Danos AM, Johnson DN, Brady MJ, Cohen RN (2009) Altering PPARgamma ligand selectivity impairs adipogenesis by thiazolidinediones but not hormonal inducers. Obesity (Silver Spring) 17(5):965–972. doi:10.1038/oby.2008.629
8. Bayindir I, Babaeikelishomi R, Kocanova S, Sousa IS, Lerch S, Hardt O, Wild S, Bosio A, Bystricky K, Herzig S, Vegiopoulos A (2015) Transcriptional pathways in cPGI2-induced adipocyte progenitor activation for browning. Front Endocrinol(Lausanne) 6:129. doi:10.3389/fendo.2015.00129
9. Rodeheffer MS, Birsoy K, Friedman JM (2008) Identification of white adipocyte progenitor

cells in vivo. Cell 135(2):240–249. doi:10.1016/j.cell.2008.09.036

10. Wu J, Bostrom P, Sparks LM, Ye L, Choi JH, Giang AH, Khandekar M, Virtanen KA, Nuutila P, Schaart G, Huang K, Tu H, van Marken Lichtenbelt WD, Hoeks J, Enerback S, Schrauwen P, Spiegelman BM (2012) Beige adipocytes are a distinct type of thermogenic fat cell in mouse and human. Cell 150(2):366–376. doi:10.1016/j.cell.2012.05.016
11. Lee YH, Petkova AP, Mottillo EP, Granneman JG (2012) In vivo identification of bipotential adipocyte progenitors recruited by beta3-adrenoceptor activation and high-fat feeding. Cell Metab 15(4):480–491. doi:10.1016/j.cmet.2012.03.009
12. Vishvanath L, MacPherson KA, Hepler C, Wang QA, Shao M, Spurgin SB, Wang MY, Kusminski CM, Morley TS, Gupta RK (2015) Pdgfrbeta mural preadipocytes contribute to adipocyte hyperplasia induced by high-fat-diet feeding and prolonged cold exposure in adult mice. Cell Metab. doi:10.1016/j.cmet.2015.10.018
13. Madsen L, Pedersen LM, Lillefosse HH, Fjaere E, Bronstad I, Hao Q, Petersen RK, Hallenborg P, Ma T, De Matteis R, Araujo P, Mercader J, Bonet ML, Hansen JB, Cannon B, Nedergaard J, Wang J, Cinti S, Voshol P, Doskeland SO, Kristiansen K (2010) UCP1 induction during recruitment of brown adipocytes in white adipose tissue is dependent on cyclooxygenase activity. PLoS One 5(6):e11391. doi:10.1371/journal.pone.0011391
14. Vegiopoulos A, Muller-Decker K, Strzoda D, Schmitt I, Chichelnitskiy E, Ostertag A, Berriel Diaz M, Rozman J, Hrabe de Angelis M, Nusing RM, Meyer CW, Wahli W, Klingenspor M, Herzig S (2010) Cyclooxygenase-2 controls energy homeostasis in mice by de novo recruitment of brown adipocytes. Science 328(5982):1158–1161. doi:10.1126/science.1186034
15. Brun RP, Tontonoz P, Forman BM, Ellis R, Chen J, Evans RM, Spiegelman BM (1996) Differential activation of adipogenesis by multiple PPAR isoforms. Genes Dev 10(8):974–984
16. Narumiya S, Sugimoto Y, Ushikubi F (1999) Prostanoid receptors: structures, properties, and functions. Physiol Rev 79(4):1193–1226
17. Ghandour RA, Giroud M, Vegiopoulos A, Herzig S, Ailhaud G, Amri EZ, Pisani DF (2016) IP-receptor and PPARs trigger the conversion of human white to brite adipocyte induced by carbaprostacyclin. Biochim Biophys Acta. doi:10.1016/j.bbalip.2016.01.007

Chapter 6

Isolation of Immune Cells from Adipose Tissue for Flow Cytometry

Jonathan R. Brestoff

Abstract

Beige and brown adipocytes are thermogenic cells essential for maintaining metabolic homeostasis within white adipose tissue (WAT) and brown adipose tissue (BAT), respectively. Emerging studies indicate that various immune cell types such as alternatively activated macrophages (AAMacs), eosinophils, and group 2 innate lymphoid cells (ILC2s) critically regulate beige and/or brown adipocyte development and activation to protect against obesity and maintain core body temperature. These findings suggest that studies of beige and brown adipose tissue may benefit from traditional immunologic approaches such as flow cytometry of immune cells residing within WAT and BAT. The purpose of this article is to describe an efficient method to isolate immune cells from numerous adipose tissue samples in parallel. The composition, phenotype, and activation state of cells isolated with this protocol may then be assessed by multiple methods including but not limited to flow cytometry. As an example, this article briefly describes a method to identify AAMacs, eosinophils, and ILC2s in adipose tissues.

Key words Beige adipocytes, Brown adipocytes, White adipose tissue, Brown adipose tissue, Alternatively activated macrophages, Eosinophils, Group 2 innate lymphoid cells

1 Introduction

Beige and brown adipocytes are specialized cells in white adipose tissue (WAT) and brown adipose tissue (BAT), respectively, that are critical for regulating adaptive thermogenesis, metabolic rate, and other critical physiologic processes [1–6]. Dysregulation of beige and brown adipocytes has been associated with a variety of metabolic diseases including but not limited to obesity and type 2 diabetes [1–6], findings that have helped to stimulate great interest in how these cells are regulated. Numerous factors have been shown to influence beige and brown adipocyte responses, and recent studies indicate that some immune cells in WAT and BAT produce effector molecules to regulate the development and activation of beige and brown adipocytes, a topic that has been recently reviewed [7, 8].

Jun Wu (ed.), *Thermogenic Fat: Methods and Protocols*, Methods in Molecular Biology, vol. 1566,
DOI 10.1007/978-1-4939-6820-6_6,

The first report of immune cell regulation of adaptive thermogenesis indicated that alternatively activated macrophages (AAMacs) in BAT produce catecholamines such as norepinephrine (NE) to defend core body temperature during short-term exposure to cold environmental temperatures [9]. AAMacs were also subsequently shown to produce NE in WAT to promote beige adipocyte responses during long-term exposure to cold temperatures [10]. In WAT, this AAMac response is controlled at least in part by eosinophils that produce interleukin (IL)-4, a type 2 cytokine that is essential for AAMac polarization and that has also been shown to directly promote beige adipocyte differentiation [10, 11]. In addition, WAT contains a recently characterized immune cell population known as group 2 innate lymphoid cells (ILC2s) that produce the type 2 cytokines IL-5 and IL-13 to sustain eosinophils and AAMacs, respectively [12]. Although ILC2s may influence beige fat responses indirectly by regulating the eosinophil/IL-4/AAMac axis, ILC2s also directly influence beige adipocyte differentiation and activation by producing IL-13 and methionine-enkephalin peptides [13, 14]. ILC2s have also been identified in BAT [13], but their functions in this tissue are not known. Collectively, these studies indicate that AAMacs, eosinophils, and ILC2s are present in WAT and BAT, where they employ multiple mechanisms to control thermogenic adipocytes.

One implication of these findings is that future studies on beige and brown adipocyte responses may benefit from characterizing the immune cell composition of adipose. Multi-color flow cytometry is a powerful tool that enables simultaneous characterization of multiple immune cell populations in terms of frequencies (relative abundance), numbers (absolute abundance), phenotypic changes, and functional potential. The purpose of this article is to describe in detail one method for isolating both common and rare immune cells from adipose tissues and for staining these cells for flow cytometric analyses. This method is suitable for one individual to process up to 24–36 adipose tissue samples in parallel, although factors that influence work flow and efficiency may increase or decrease the number of samples that can be processed using this technique. Staining panel design and flow cytometer operation are not discussed because these topics depend on the model and configuration of the specific flow cytometer that will be used. However, a small collection of articles that define staining panels and/or gating strategies to identify AAMacs, eosinophils, and ILC2s in murine adipose tissues are listed in Table 1, and basic flow cytometry resources are listed in Table 2.

Table 1
Identification of immune cells that regulate beige and/or brown adipocytes

Cell type	Basic identification in adipose[a]	References
Eosinophils	$CD45^+$ $CD11c^-$ $SiglecF^+$ SSC^{hi}	[11]
Alternatively activated macrophages (AAMacs)[b]	$CD45^+$ $SiglecF^-$ $F4/80^+$ $CD11b^+$ $CD206^+$ ($YARG^+$ if using YARG reporter mice)	[9, 12]
Group 2 innate lymphoid cells (ILC2s)[c]	$CD45^{int}$ $Lineage^-$ $CD25^+$ $IL\text{-}33R^+$	[12, 13, 15]

[a]These basic approaches to identify the indicated cell types are summarized from the listed peer-reviewed articles. These approaches are sufficient for cell identification in most cases but do not enable comprehensive phenotypic characterization. For gating strategies, see the referenced articles.
[b]AAMacs from tissue are notoriously difficult to stain and characterize by flow cytometry. CD206 should be used as an intracellular stain. Enhanced yellow fluorescent protein-Arginase 1 (YARG) transgenic reporter mice are useful for identifying Arginase 1-expressing AAMacs.
[c]CD45 expression on ILC2s is intermediate (^{int}). $Lineage^-$ indicates the absence of the following lineage markers: CD3, CD5, CD19, B220, NK1.1, CD11c, CD11b, SiglecF, and FcεRIα.

Table 2
Online flow cytometry resources

Resource	Notes
ThermoFisher Scientific Flow Cytometry Resource Center	Flow cytometry basics Analyzing flow cytometry data Flow cytometry protocols Numerous guides and webinars Information about relevant products
Affymetrix eBioscience flow cytometry website	Flow cytometry technical guide Flow cytometry protocols Fluorophore charts Links to human and mouse cross-reactivity charts
BD Biosciences multicolor flow cytometry website	Basic online introduction to flow cytometry Numerous protocols, guides, and webinars Helpful tools: fluorescence spectrum viewer, absorption and emission spectra, buffer compatibility charts, panel design tool

2 Materials

This protocol involves handling of murine adipose tissues and primary cells; therefore, all solutions and reagents should be prepared under sterile conditions. Generally, reagents are prepared using sterile phosphate buffered saline (PBS) at pH 7.4 to ensure maintenance of iso-osmotic conditions. Use of water is not recommended except for preparation of iso-osmotic reagents, and in such instances it is critical to use sterile distilled, deionized water. Reagents should be stored at 4 °C with protection from light. General and institution-specific laboratory precautions and waste disposal regulations need to be followed diligently to safely use this protocol.

2.1 Isolation of Immune Cells from Adipose Tissue

1. Phosphate Buffered Saline (PBS): Obtain 1.0 L of sterile PBS.
2. Fetal bovine serum (FBS): Thaw 0.5–1.0 L of sterile FBS and heat-inactivate at 55 °C for 1 h using a water bath or bead bath. Divide the heat-inactivated FBS into 50 mL aliquots and store at −20 °C until use.
3. Dulbecco's Modified Eagle Medium (DMEM): High glucose (4.5 g/L) with calcium, without sodium pyruvate. Store at 4 °C.
4. Collagenase Type II: 100 mg/mL in PBS (100× stock). Filter-sterilize using a 0.45 μm filter system and divide into 0.25 mL aliquots. Store at −20 °C until use (*see* **Note 1**).
5. 100× penicillin-streptomycin-glutamine: 10,000 units penicillin G, 10,000 μg streptomycin sulfate, and 200 mM L-glutamine in a buffer of 10 mM citrate. Divide into 10 mL aliquots and store at −20 °C until use.
6. Shaker: Set at 37 °C and 150–200 rotations per minute. The shaker platform must be equipped with a rack that can accommodate 50 mL conical tubes positioned securely at a 45° angle. The rotation rate may vary depending on the shaker's orbit.
7. Sharp dissecting scissors: autoclaved.
8. Bijou sample containers: 7 mL volume capacity. One bijou container per sample.
9. 50 mL conical tubes: Two 50 mL conical tubes per sample.
10. Cell strainers: 70–100 μm mesh size, sterile. One strainer per sample.
11. Sterile 3 mL syringes.
12. Motorized pipette filler with graduated 10 and 25 mL serologic pipettes (sterile).
13. Plastic or foam tube racks that can securely accommodate 50 mL conical tubes.
14. ACK Red Blood Cell Lysis Buffer.
15. Low-speed centrifuge outfitted with a swinging bucket rotor and adapters for 50 mL conical tubes and tissue culture plates. All centrifugation steps will be performed using the following settings: 500 × *g* for 5 min at 4 °C with fast acceleration and braked deceleration.

2.2 Staining Immune Cells for Flow Cytometry

1. FACS Buffer: Add 10 g of bovine serum albumin and 1 mL of 500 mM EDTA to 1 L of PBS.
2. Blocking Buffer (prepare fresh): Add 20 μL of 0.5 mg/mL Rat IgG (final concentration of 10 μg/mL) and 20 μL of 0.5 mg/mL anti-CD16/32 (clone 2.4G2, final concentration of 10 μg/mL) to each 960 μL FACS Buffer required. Make 150 μL of Blocking Buffer for each sample.
3. Fixable Viability Stain (prepare fresh): Add 1 μL of LIVE/DEAD® Fixable Dead Cell Stain (ThermoFisher Scientific) to each 600 μL of PBS. Make 100 μL for each sample. An alternative viability stain may be used.
4. Primary Stain (prepare fresh): Add selected fluorophore-conjugated and biotin-conjugated anti-mouse antibodies to FACS Buffer according to dilution factors obtained from the manufacturer, primary literature, or empiric testing. An unconjugated anti-mouse antibody can be attempted if it is generated in a species not used to generate any of the other antibodies. Make 100 μL of Primary Stain for each sample (*see* **Note 2**).
5. Secondary Stain (prepare fresh): Add fluorophore-conjugated biotin and/or fluorophore-conjugated species-specific antibody (e.g., goat anti-rabbit IgG) to FACS Buffer according to dilution factors obtained from the manufacturer, primary literature, or empiric testing.

2.3 Intracellular Staining of Immune Cells for Flow Cytometry

1. Fixation/Permeabilization Buffer Set (prepare fresh): Select a suitable fixation/permeabilization buffer set such as the Foxp3 Staining Kit (eBioscience), Cytofix Fixation Buffer (BD), or an alternative product. Make 100 μL for each sample according to manufacturer's instructions (*see* **Note 3**).
2. Permeabilization Buffer: Mix 45 mL of distilled, deionized water, and 5 mL of 10× Permeabilization Buffer (eBioscience, catalog number 00-8333-56) to obtain a 1×working solution, which can be stored at 4 °C for at least 2 days.
3. Intracellular Stain: Add antibodies against intracellular antigens to 1× Permeabilization Buffer using desired dilution factors. Make 100 μL for each sample.

2.4 Cell Fixation for Delayed Flow Cytometry

1. Fixation Buffer: 2% paraformaldehyde in PBS.

2.5 Final Cell Preparation for Flow Cytometry

1. Filter-top round-bottom FACS tubes: one per sample.
2. Cell counting beads (*see* **Note 4**).

3 Methods

3.1 Isolation of Immune Cells from Adipose Tissue

1. Prepare fresh Wash Media: Remove 60 mL of DMEM from a 1 L bottle of DMEM and set aside for the preparation of Digestion Media. To the remaining 940 mL of DMEM, add 50 mL of heat-inactivated FBS (final concentration: 5%) and 10 mL of 100× penicillin–streptomycin–glutamine (final concentration: 1×).
2. Prepare fresh Digestion Media: Add 1 volume of stock Collagenase Type II (100 mg/mL) to 99 volumes of DMEM to obtain a working concentration of 1 mg/mL Collagenase Type II in DMEM (e.g., add 500 μL of stock Collagenase Type II to 49.5 mL of DMEM). In bulk, prepare 3 mL Digestion Media for each sample.
3. Add 3 mL Digestion Media to each bijou container (one container for each adipose sample). Recap the containers, and weigh each bijou containing Digestion Media (Weight #1).
4. According to a protocol approved for use by the investigator's Institutional Animal Care and Use Committee or equivalent ethics committee, euthanize the first experimental animal.
5. Rapidly but carefully dissect the desired adipose tissue depots and immediately place the harvested tissues in the bijou containing Digestion Media (*see* **Note 5**).
6. Weigh the bijou containing Digestion Media and adipose tissue (Weight #2). Subtract Weight #1 from Weight #2 to determine the amount of adipose being utilized (*see* **Note 6**).
7. Place the bijou with Digestion Media and adipose tissue on ice and proceed to harvest tissue from the next animal.
8. Repeat **steps 5**–**7** until the tissue harvest is complete.
9. Finely mince each adipose tissue sample in the Digestion Media to <1 mm pieces using sharp dissecting scissors (*see* **Note 7**).
10. Place each bijou container in a 50 mL conical tube for secondary containment (*see* **Note 8**).
11. Incubate the samples in a shaker at 37 °C with 150–200 rotations per minute for 45–60 min with the samples positioned at a 45° angle (*see* **Note 9**).
12. Place cell strainers into new 50 mL conical tubes and pour the digested tissues into the cell strainer basins. Collect the flow-through and, if undigested adipose tissue pieces remain on the mesh surface, use the plunger of a 3 mL syringe to gently attempt to dissociate the tissue.
13. To maximize cell yield and quench the digestion reaction, rinse each bijou with approximately 5 mL Wash Media and add the rinse to the cell strainer to collect the flow-through.
14. Add 5 mL Wash Media to each cell strainer and collect the flow-through. Repeat.

15. Centrifuge the 50 mL conical tubes for 5 min at 500 × *g* at 4 °C.
16. Immediately after centrifugation, carefully remove the tubes and aspirate the floating adipocytes and supernatant, taking care not to disrupt the cell pellet (*see* **Note 10**).
17. Add 0.5 mL ACK Red Blood Cell Lysis Buffer (room temperature) to each cell pellet, mix, and incubate at room temperature for 2–5 min.
18. Add 10–15 mL Wash Media and centrifuge for 5 min at 500 × *g* at 4 °C. Aspirate the supernatants as above.
19. Resuspend each cell pellet in 150 μL Wash Media and transfer the entirety of each sample to individual wells of a round-bottom or V-bottom 96-well plate. Proceed to Subheading 3.2 (*see* **Note 11**).

3.2 Staining Immune Cells for Flow Cytometry

1. Centrifuge the plate containing cells for 5 min at 500 × *g* at 4 °C and discard the supernatants.
2. Resuspend each sample in 200 μL of PBS, centrifuge for 5 min at 500 × *g* at 4 °C, and discard the supernatants.
3. Resuspend the cells in 100 μL of Fixable Viability Stain and incubate on ice for 30 min or according to manufacturer instructions.
4. Add 100 μL of FACS Buffer to bring to a final volume of 200 μL, and then centrifuge for 5 min at 500 × *g* at 4 °C. Discard the supernatants.
5. Resuspend the cells in 200 μL of FACS Buffer, centrifuge for 5 min at 500 × *g* at 4 °C, and discard the supernatants.
6. Resuspend the cells in 100 μL of Blocking Buffer and incubating on ice for 15 min to minimize nonspecific binding. Add 100 μL of FACS Buffer to bring to a final volume of 200 μL, and then centrifuge for 5 min at 500 × *g* at 4 °C. Discard the supernatants (*see* **Note 12**).
7. Resuspend the cells in 100 μL of Primary Stain, incubate on ice for 30 min, centrifuge for 5 min at 500 × *g* at 4 °C and discard the supernatants.
8. Resuspend the cells in 200 μL of FACS Buffer to wash the cells, centrifuge for 5 min at 500 × *g* at 4 °C, and discard the supernatants. Repeat two additional times (three washes total).
9. If necessary based on the staining panel, resuspend the cells in 100 μL of Secondary Stain, incubate on ice for 30 min, centrifuge for 5 min at 500 × *g* at 4 °C, and discard the supernatants.
10. Repeat **step 8**.
11. Proceed to Subheading 3.3 for intracellular staining (if necessary), Subheading 3.4 for cell fixation for delayed flow cytometry (if necessary), or Subheading 3.5 for final cell preparation for flow cytometry.

3.3 Intracellular Staining of Immune Cells for Flow Cytometry

1. Resuspend the cells in Fixation/Permeation Buffer and incubate according to the manufacturer's instructions (*see* **Note 3**).
2. Centrifuge for 5 min at 500 × *g* at 4 °C, discard the supernatants, and resuspend the cells in 200 μL of 1×Permeabilization Buffer. Repeat two additional times (three washes total).
3. At the end of the third wash, resuspend the cells in 100 μL of Intracellular Stain containing antibodies against intracellular epitopes and incubate on ice for 30–60 min (*see* **Notes 13** and **14**).
4. Repeat **step 2**.
5. At the end of the third wash, resuspend the cells in 200 μL of FACS Buffer, centrifuge for 5 min at 500 × *g* at 4 °C, and discard the supernatants.
6. Resuspend the cells in 150 μL of FACS Buffer and proceed to Subheading 3.5 for Final Cell Preparation for Flow Cytometry.

3.4 Cell Fixation for Delayed Flow Cytometry

1. Resuspend live, surface-stained cells in 150 μL of Fixation Buffer and incubate for 20 min at 4 °C.
2. Centrifuge for 5 min at 500 × *g* at 4 °C and discard the supernatants.
3. Resuspend the cells in 200 μL of FACS Buffer for storage overnight at 4 °C (*see* **Note 15**).
4. The next day centrifuge the plate for 5 min at 500 × *g* at 4 °C and discard the supernatants. Proceed to Subheading 3.5 for Final Cell Preparation for Flow Cytometry.

3.5 Final Cell Preparation for Flow Cytometry

1. Resuspend the cells in 150 μL of FACS Buffer and transfer the entirety of each sample into the filter-top of its own filter-top round-bottom flow cytometer tube.
2. Collect all of the cells and FACS Buffer through the filter by pulse centrifugation at 500 × *g* for 10–15 s.
3. Remove the filter-tops and add equal volumes of cell counting beads to each sample, according to manufacturer's instructions. Mix.
4. Run the samples on a flow cytometer (for resources on flow cytometry, *see* Table 2). Acquire data for the entire sample for best results.

4 Notes

1. Collagenase Type II enables excellent digestion of WAT. This enzyme is effective but less efficient for BAT digestion.
2. Most anti-mouse antibodies are generated in rats. The source species is usually identified in the product information sheet and can be located online.

3. There are several commercially available Fixation/Permeabilization Buffer Sets. The Foxp3 Staining Kit (eBioscience) is useful for nuclear stains and most effector cytokines, and BD Cytofix Fixation Buffer is useful for most cytoplasmic stains and most effector cytokines. Other fixation and permeabilization buffers are available or can be made. Selection of the ideal fixation/permeabilization buffer set is empiric.
4. When selecting counting beads and designing a staining panel, check to ensure that the excitation and emission spectra are unlikely to create a compensation problem or interfere with other fluorophores being used.
5. For epididymal and ovarian WAT, avoid harvesting reproductive structures. For inguinal WAT, use a punch biopsy tool to remove the entire inguinal lymph node before subjecting the inguinal adipose to mechanical or enzymatic digestion.
6. The tissue weights are used to express flow cytometry data as the number of cells per gram of adipose.
7. Adequate mincing is achieved when pieces are approximately 1 mm or less. Larger pieces do not digest efficiently.
8. If there is a leak, the 50 mL conical tube used for secondary containment will collect the fluid to prevent loss of sample.
9. The digestion should be monitored visually every 15 min starting at the 30 min time point. Digestion is complete when the Digestion Media appears milky with few or no tissue pieces remaining. If the digestion is incomplete after 60 min, the incubation can be extended up to 90 min without substantially compromising overall cell viability in the short-term (i.e., for immediate processing and staining for flow cytometry). Longer digestion times under these conditions may not be suitable for all applications, and cell viability or phenotypic markers of interest may need to be evaluated on a case-by-case basis.
10. After centrifugation, do not allow the samples to sit in the centrifuge. The small cell pellet is easily disrupted, and exposing the pellet to vibrations from the centrifuge's refrigeration unit may lead to significant cell losses. This precaution applies throughout the protocol.
11. Counting cells in a hemocytometer is not feasible for most adipose samples because total cell yield from a single gonadal fat pad obtained from a lean, healthy mouse is likely to be less than 100,000 cells depending on the age and sex of the animal and the amount of tissue collected. A cell yield this low is not reliably counted on a hemocytometer without sacrificing large proportions of the sample, which may compromise the quality of flow cytometry plots, especially for rare cell populations.
12. If anti-CD16/32 is part of the staining panel, do not expose the cells to Blocking Buffer. Instead, first staincells with the

fluorophore-conjugated anti-CD16/32 antibody. Then wash and stain with the remainder of the staining panel.

13. Intracellular staining is useful for cytokines, nuclear proteins such as transcription factors, or other molecules with intracellular epitopes.
14. Since the cells are fixed, the stain duration can be extended overnight, however an overnight intracellular stain might increase non-specific binding.
15. Fluorophores may degrade over time or with fixation. The adequacy of a fluorophore's signal after fixation or overnight storage can be determined on an empiric basis.

Acknowledgments

This work was supported by NIH grant AI112023.

References

1. Cannon B, Nedergaard J (2004) Brown adipose tissue: Function and physiological significance. Physiol Rev 84(1):277–359. doi:10.1152/physrev.00015.2003
2. Harms M, Seale P (2013) Brown and beige fat: development, function and therapeutic potential. Nat Med 19(10):1252–1263. doi:10.1038/nm.3361
3. Pfeifer A, Hoffmann LS (2014) Brown, beige, and white: the new color code of fat and its pharmacological implications. Annu Rev Pharmacol Toxicol. doi:10.1146/annurev-pharmtox-010814-124346
4. Wu J, Cohen P, Spiegelman BM (2013) Adaptive thermogenesis in adipocytes: is beige the new brown? Genes Dev 27(3):234–250. doi:10.1101/gad.211649.112
5. Cohen P, Levy JD, Zhang Y, Frontini A, Kolodin DP, Svensson KJ, Lo JC, Zeng X, Ye L, Khandekar MJ, Wu J, Gunawardana SC, Banks AS, Camporez JP, Jurczak MJ, Kajimura S, Piston DW, Mathis D, Cinti S, Shulman GI, Seale P, Spiegelman BM (2014) Ablation of PRDM16 and beige adipose causes metabolic dysfunction and a subcutaneous to visceral fat switch. Cell 156(1–2):304–316. doi:10.1016/j.cell.2013.12.021
6. Wu J, Bostrom P, Sparks LM, Ye L, Choi JH, Giang AH, Khandekar M, Virtanen KA, Nuutila P, Schaart G, Huang K, Tu H, van Marken Lichtenbelt WD, Hoeks J, Enerback S, Schrauwen P, Spiegelman BM (2012) Beige adipocytes are a distinct type of thermogenic fat cell in mouse and human. Cell 150(2):366–376. doi:10.1016/j.cell.2012.05.016
7. Brestoff JR, Artis D (2015) Immune regulation of metabolic homeostasis in health and disease. Cell 161(1):146–160. doi:10.1016/j.cell.2015.02.022
8. DiSpirito JR, Mathis D (2015) Immunological contributions to adipose tissue homeostasis. Semin Immunol 27(5):315–321. doi:10.1016/j.smim.2015.10.005
9. Nguyen KD, Qiu Y, Cui X, Goh YP, Mwangi J, David T, Mukundan L, Brombacher F, Locksley RM, Chawla A (2011) Alternatively activated macrophages produce catecholamines to sustain adaptive thermogenesis. Nature 480(7375):104–108. doi:10.1038/nature10653
10. Qiu Y, Nguyen KD, Odegaard JI, Cui X, Tian X, Locksley RM, Palmiter RD, Chawla A (2014) Eosinophils and type 2 cytokine signaling in macrophages orchestrate development of functional beige fat. Cell 157(6):1292–1308. doi:10.1016/j.cell.2014.03.066
11. Wu D, Molofsky AB, Liang HE, Ricardo-Gonzalez RR, Jouihan HA, Bando JK, Chawla A, Locksley RM (2011) Eosinophils sustain adipose alternatively activated macrophages associated with glucose homeostasis. Science 332(6026):243–247. doi:10.1126/science.1201475
12. Molofsky AB, Nussbaum JC, Liang HE, Van Dyken SJ, Cheng LE, Mohapatra A, Chawla A, Locksley RM (2013) Innate lymphoid type 2 cells sustain visceral adipose tissue eosinophils and alternatively activated macrophages. J Exp Med 210(3):535–549. doi:10.1084/jem.20121964

13. Brestoff JR, Kim BS, Saenz SA, Stine RR, Monticelli LA, Sonnenberg GF, Thome JJ, Farber DL, Lutfy K, Seale P, Artis D (2015) Group 2 innate lymphoid cells promote beiging of white adipose tissue and limit obesity. Nature 519:242–246. doi:10.1038/nature14115

14. Lee MW, Odegaard JI, Mukundan L, Qiu Y, Molofsky AB, Nussbaum JC, Yun K, Locksley RM, Chawla A (2015) Activated type 2 innate lymphoid cells regulate beige fat biogenesis. Cell 160(1–2):74–87.doi:10.1016/j.cell.2014.12.011

15. Hams E, Locksley RM, McKenzie AN, Fallon PG (2013) Cutting edge: IL-25 elicits innate lymphoid type 2 and type II NKT cells that regulate obesity in mice. J Immunol 191(11):5349–5353. doi:10.4049/jimmunol.1301176

Chapter 7

Differentiation and Metabolic Interrogation of Human Adipocytes

Nicki A. Baker, Lindsey A. Muir, Carey N. Lumeng, and Robert W. O'Rourke

Abstract

Adipocytes differentiated from preadipocytes provide a valuable model for the study of human adipocyte metabolism. We describe methods for isolation of human stromal vascular cells, expansion of preadipocytes, differentiation into mature adipocytes, and in vitro metabolic interrogation of adipocytes.

Key words Adipocyte, Preadipocyte, Stromal vascular cell fraction, Adipose tissue, Differentiation, Lipogenesis, Lipolysis, Glucose uptake

Abbreviations

eWAT	Inguinal white adipose tissue (iWAT)
PA	Preadipocyte
PBS	Phosphate buffered saline
QRTPCR	Quantitative real-time polymerase chain reaction
SAT	Subcutaneous adipose tissue
SVF	Stromal-vascular cell fraction
VAT	Visceral adipose tissue

1 Introduction

Obesity is a public health crisis. Adipose tissue metabolic dysfunction is a central feature of obesity and underlies the pathogenesis of metabolic disease. An understanding of adipocyte metabolism is therefore of critical importance to metabolic disease research. Murine 3T3L1 cells provide a valuable adipocyte model, but significant differences exist between murine and human adipose tissue and systemic metabolic disease phenotypes. Human adipocytes differentiated *in vitro* from preadipocytes retain depot- and

Jun Wu (ed.), *Thermogenic Fat: Methods and Protocols*, Methods in Molecular Biology, vol. 1566,
DOI 10.1007/978-1-4939-6820-6_7,

patient-specific metabolic phenotypes in culture and provide a tractable model for studying human adipocyte cellular metabolism [1–5].

While contributing to the vast majority of total adipose tissue mass, mature adipocytes comprise only half of total adipose tissue cell number, the remainder of which consists of a diverse stromal vascular cell fraction (SVF). Forty to sixty percent of the SVF is CD45+ leukocytes including macrophages, T-cells, B-cells, and other immune cells. Of the remaining CD45− SVF population, over 80 % are CD34+CD31-, a phenotype that defines preadipocytes (a.k.a. adipocyte stem cells), a mesenchymal cell population that gives rise to mature adipocytes. Endothelial cells, fibroblasts, and other cell types comprise the remainder of the CD45−population; endothelial cells are typically described as CD45-CD34+CD31+, and comprise 1–5 % of the CD45− SVF population (Fig. 1). Other markers define specific subpopulations of human preadipocytes with variable adipogenic capacity. CD140a+ preadipocytes, for example, are a particularly adipogenic subset of the CD34+ CD31− population [6, 7]. As of this writing, the functional significance of specific human preadipocyte subpopulations is poorly defined and an active area of research.

Preadipocytes may be isolated or enriched from SVF using a variety of methods depending on the intended downstream application. Flow cytometry sorting of CD45-CD34+CD31+ preadipocytes provides a high purity cell preparation for microarray and next-generation sequencing studies; antibody-coated magnetic bead sorting also provides a relatively high purity population. Enrichment of preadipocytes by plastic adherence of SVF provides

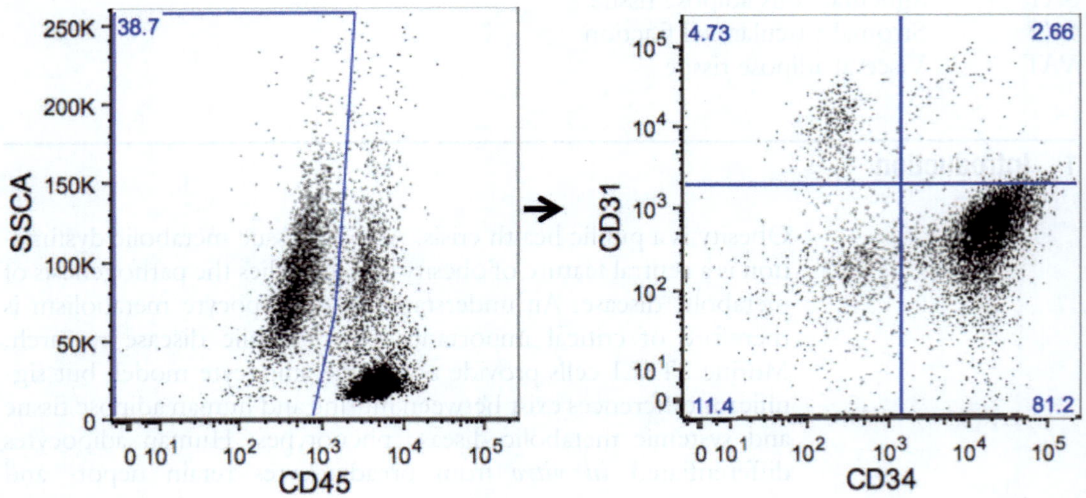

Fig. 1 Flow cytometry of human SVF: Representative flow cytometry scatter plots of visceral human adipose tissue SVF demonstrating gating strategy for preadipocytes (CD45-CD31-CD34+); a large forward scatter-side scatter gate is used to encompass all viable cells (not shown), followed by gating on CD45- cells (*left*), followed by gating on CD31-CD34+ cells (*right, right lower quadrant*)

a cell population of sufficient purity for many applications and is efficient and economical. Nonadherent CD45+ hematopoietic cells are removed after 1–2 days of adherence, and CD45-CD34+CD31- cells subsequently selectively proliferate in culture and are receptive to adipogenic differentiation.

Adipose tissue may be divided into white and brown phenotypes, although increasing data suggest functional overlap, with pluripotent stem cells giving rise to white, brown, and beige/brite (brown-in-white) adipocytes. While white and brown adipose tissues are for the most part anatomically distinct, stem cells with brown/beige potential have been identified within white anatomic depots [8], reinforcing functional overlap between canonical adipose tissue depots. In humans, the existence of brown adipose tissue beyond the neonatal period was confirmed using positron emission tomography only in the last decade [9]. As a result, most published literature prior to this point studies white adipose tissue.

White adipose tissue in humans and mice resides in anatomically and functionally distinct depots that may be broadly categorized into visceral and subcutaneous adipose tissue compartments (VAT, SAT). Multiple sub-depots within each compartment have distinct phenotypes and functions. Adipose tissue depot anatomy differs significantly in mice and humans. The epididymal fat pad comprises the majority of VAT in mice, is often termed eWAT (epididymal white adipose tissue), and is the best-studied murine VAT depot, although mice also harbor mesenteric and retroperitoneal VAT. Murine SAT is generally derived from the inguinal and flank areas (iWAT, inguinal white adipose tissue). As in humans, murine VAT/eWAT, when compared to SAT/iWAT, is more strongly associated with systemic metabolic disease [10–12]. VAT in humans includes omental, mesenteric, and retroperitoneal depots. The greater omentum tissue consists of an apron of adipose tissue attached to the transverse colon and greater curvature of the stomach, and is the most easily accessible and best studied human VAT sub-depot (Fig. 2). Human SAT may be divided into truncal (abdominal wall) and extremity compartments, within which exist deep and superficial sub-depots.

Technical and anatomic considerations affect surgical adipose tissue biopsy in humans. Omental adipose tissue is easily accessible during most intra-abdominal operations and may be biopsied without significant risk or clinical sequelae using cautery or harmonic scalpel dissection. Mesenteric adipose tissue is intimately associated with the small intestine, and thus biopsy is usually performed only in the context of clinically indicated small bowel resection. Retroperitoneal VAT is intimately associated with the kidneys and other retroperitoneal structures; retroperitoneal adipose tissue biopsy is therefore only possible in the context of clinically indicated resection of the kidney or other retroperitoneal organs. SAT biopsy is usually performed at the site of the skin incision for most operations, the anatomic location of which varies depending on the specific operation performed. We obtain omental VAT and

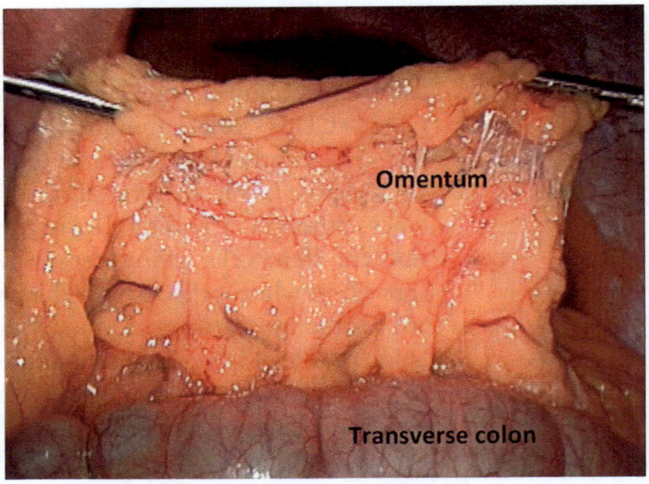

Fig. 2 The human omentum: Intraoperative photograph of the human omentum reflected ventrally off the transverse colon; the view is from a caudad position looking cephalad, with ventral and dorsal positions at top and bottom of photo respectively; the stomach (not seen) is cephalad, behind the omentum in the photograph

abdominal wall SAT from patients undergoing bariatric surgery and other abdominal operations via laparoscopy and laparotomy. Biopsy of these tissues via laparotomy is relatively straightforward for the surgeon. Omental VAT collection via laparoscopy is technically more challenging and time-consuming, and requires intraoperative extraction from the abdomen through a laparoscopic trocar site with a laparoscopic specimen retrieval bag, removing tissue from the bag intraoperatively through the trocar site using ringed forceps. We endeavor to collect VAT and SAT from the same anatomic sites in all patients whenever possible. VAT is collected from the apex/caudad end of the omentum. Since the majority of samples are collected from laparoscopic bariatric operations, we standardize SAT collection to the upper abdominal wall at the site of a specific trocar incision placed in every operation. We collect 20–80 g of VAT from the greater omentum, and 1–6 g of SAT from the skin incision. SAT samples larger than 6 g from a laparotomy incision, or larger than 4 g from a laparoscopy trocar site, are associated with higher rates of hematoma, seroma, and infection, and are contraindicated. Biopsy size, especially SAT, should be tailored to the amount of adipose tissue present, at the discretion of the operating surgeon. Each laboratory should work closely with surgeon investigators/collaborators to refine and standardize surgical technique.

Tissue is transported to the laboratory in a sterile plastic bag on ice and samples aliquoted and stored for various applications using a codified standard operating protocol. Here, we provide protocols for isolation of SVF from human adipose tissue, expansion of preadipocytes from SVF, and differentiation of preadipocytes into mature adipocytes, but whole tissue may be frozen, formalin-fixed, or cultured as live explants, and leukocyte and non-leukocyte cell subpopulations may be isolated from SVF using flow cytometry or antibody-coated magnetic bead sorting for multiple phenotypic and functional analyses. Our protocols have been optimized for omental VAT and abdominal wall SAT, but have also been successfully used for retroperitoneal and mesenteric tissues. Further optimization for different sub-depots may improve results and should be considered by individual laboratories.

2 Materials

2.1 Equipment

1. Laminar flow biosafety cabinet.
2. Temperature-controlled orbital shaker.
3. Temperature-controlled desktop centrifuge.
4. Tissue culture incubator, 37 °C, 5 % CO_2.
5. Freezer, −80 °C.
6. Liquid nitrogen storage unit.
7. Microplate spectrophotometer.
8. Scintillation counter.
9. Sterile scissors, forceps.
10. Hemocytometer.

2.2 Disposables

1. Tissue culture materials, including 10 cm culture plates, T150 cell culture flasks, 24-well plates, 50 mL conical tubes.
2. 100 μM nylon cell strainers.
3. Plastic weigh boats.
4. Internally threaded cryogenic vials.
5. 1.5 mL microcentrifuge tubes.
6. Scintillation vials, scintillation fluid.

2.3 Isolation of SVF from Human Adipose Tissue

1. 20× collagenase stock solution: Type II Collagenase (Gibco Inc.) 40 mg/mL in 1× PBS, 2 % BSA; prepare immediately prior to use.
2. RBC lysis buffer: 10× stock solution: 82.9 g ammonium chloride + 10.0 g potassium bicarbonate + 0.37 g EDTA, QS to 1 liter with sterile H_2O; dilute to 1× with sterile H_2O prior to use.
3. Trypan blue.

2.4 Expansion and Storage of Preadipocytes

1. Expansion medium: DMEM/F12 (50:50), 15 % FBS, 1 % antibiotic-antimycotic.
2. Freezing medium: DMEM/F12 (50:50), 15 % FBS, 10 % DMSO, 1 % antibiotic-antimycotic.
3. Trypsin solution: 0.25 % Trypsin-EDTA.

2.5 In Vitro Differentiation of Human Adipocytes

1. Differentiation medium: Serum-free DMEM/F12 (50:50), 2.5 mM glutamine, 15 mM HEPES, 1 % antibiotic-antimycotic, 10 mg/ml transferrin, 33 μM biotin, 0.5 μM human insulin, 17 μM pantothenate, 0.1 μM dexamethasone, 2 nM T3, 540 μM IBMX, 1 μM ciglitazone.
2. Maintenance medium: Serum-free DMEM/F12 (50:50), 2.5 mM glutamine, 15 mM HEPES, 1 % antibiotic-antimycotic, 10 mg/ml transferrin, 33 μM biotin, 0.5 μM human insulin.

2.6 Oil Red-O Staining

1. Oil Red-O working solution: Add 30 mL of Oil Red-O stock solution (Sigma Inc.) to 20 mL of H_2O and filter solution through 15 cm Grade 201 Whatman-Reeve Angel filter paper.
2. Formalin, 4 %.
3. Isopropanol, 60 %.

2.7 Lipolysis Assay

1. Maintenance medium without insulin: Serum-free DMEM/F12 (50:50), 2.5 mM glutamine, 15 mM HEPES, 1 % antibiotic-antimycotic, 10 mg/ml transferrin, 33 μM biotin.
2. Triglyceride Determination Kit (Sigma Inc.).
3. Isoproterenol hydrochloride.

2.8 Labeled Acetate Lipogenesis Assay

1. Serum Starvation Medium: DMEM:F12 50:50, 1 % antibiotic-antimycotic.
2. Lipogenesis Medium: Serum Starvation Medium + 100 nM insulin, 10 μM sodium acetate, 0.5uCi ^{3}H-acetate.
3. Human insulin stock solution, 1 μM; dilute to 100 nM working solution in 1× PBS.
4. Sodium acetate 10 mM stock solution.
5. ^{3}H–acetate (Perkin Elmer Inc.).
6. HCl stock solution, 1 N; dilute to 0.1 N HCl with H_2O on the day of procedure to obtain working solution.
7. Chloroform:methanol 2:1 v/v.

3 Methods

3.1 Isolation of SVF from Human Adipose Tissue

1. Transport human adipose tissue to laboratory in a sterile plastic bag on ice. All tissue processing is done in a laminar flow biosafety cabinet under sterile tissue culture conditions. Weigh

tissue and then transfer to a sterile plastic tray. Remove and discard blood clots, vessels, and cauterized tissue with sterile scissors and forceps. Mince remaining tissue into 2–3 mm (~5 mg) pieces with scissors and forceps. Mincing should be performed rapidly, for not more than 15–20 min, to prevent decreased cell viability. Use multiple personnel to mince tissue in parallel for large samples, to keep mincing time to a minimum (*see* **Note 1a**).

2. Aliquot 2 mL (2 g) of minced adipose tissue into a 50 mL conical tube with a transfer pipette; use multiple tubes for larger samples. To each tube, add 17 mL of 1×PBS + 2 % BSA. At this point, if any remaining visible pieces of tissue >2 mm are present, then further mince tissue in buffer in the tube with scissors until tissue fragments are <2 mm.
3. Add 1 mL of 20× collagenase stock solution to the 19 mL tissue suspension to a total volume of 20 mL and a final working (1×) collagenase concentration of 2 mg/mL. Agitate samples on an orbital shaker for 30–60 min, 130 rpm, 37 °C, removing tubes every 5 min to agitate by hand more vigorously for 5–10 s. The resulting digestate should be cloudy, homogenous, and almost fully liquefied with some strands of gelatinous, fibrous tissue aggregates (*see* **Notes 1b–d**).
4. While tissue is digesting, pre-wet 100 μM nylon cell strainers with 1× PBS and place on top of fresh 50 mL conical tubes (one for each digestate tube). After digestion is complete, apply the tissue digestate from a single 50 mL tube onto a single strainer and allow it to gravity drip into the 50 mL tube; use 10 mL of 1×PBS to rinse the initial digestion 50 mL tube, removing any residual digestate, and apply to the strainer to rinse, letting it gravity drip into tube. Repeat for all digestate tubes. Rinse each strainer with an additional 10 ml 1× PBS. The strainer will remove strands of gelatinous, fibrous tissue aggregates. Discard the strainer with these fibrous tissue aggregates and retain the filtrate (*see* **Note 1e**).
5. Centrifuge the filtrate for 10 min, 250 rcf, 4 °C; the sample will layer into three fractions from bottom to top: an SVF pellet, an aqueous supernatant, and a lipid/adipocyte layer (Fig. 3). Pipette off the lipid/adipocyte and aqueous layers with a transfer pipette, being careful not to disturb the SVF pellet, leaving only the SVF pellet with 1-2 mL of overlying aqueous phase. Combine SVF pellets from three to five conical tubes into one fresh 50 mL conical tube.
6. Add 20 mL 1× RBC lysis solution to the SVF pellet; vortex to mix, and incubate 5 min, 25 °C. Dilute RBC lysate with 30 mL of 1× PBS to tube and invert tube to mix. Centrifuge 10 min, 250rcf, 4 °C. Remove supernatant from cell pellet with a pipette.

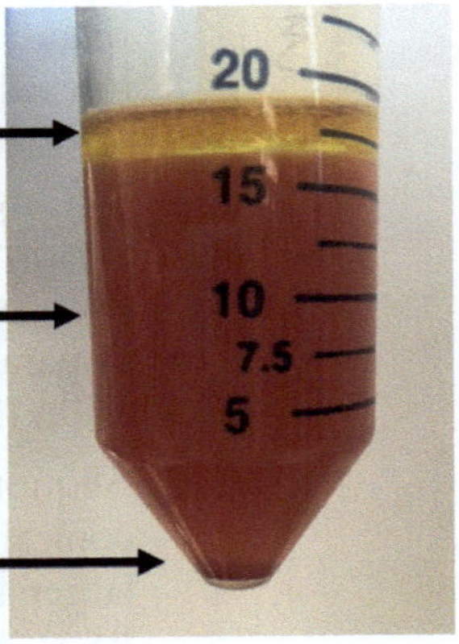

Fig. 3 Collagenase-digested human visceral adipose tissue: Collagenase-digested human visceral adipose tissue after gradient centrifugation demonstrating lipid and aqueous layers and the SVF cell pellet

7. Resuspend SVF cell pellet in either in 1×PBS or culture medium depending on intended downstream application (e.g., PBS for flow cytometry; medium for culture/differentiation) (*see* **Note 1f**). Remove a 10 μL aliquot and mix with 10 μL trypan blue in a microcentrifuge tube. Count viable SVF cells using a hemocytometer or automated cell counter.

3.2 Expansion and Storage of Preadipocytes

1. Seed 5 × 10^5 fresh SVF cells into a 10 cm cell culture dish (~1 × 10^4 cells/cm^2) (*see* **Note 1g**). Add 10 mL expansion medium. Incubate for 24 h. Place in an incubator, change medium every 2–3 days until 90 % + confluent, usually 2–3 days. Non-preadipocytes (leukocytes, endothelial cells) will not adhere and will be removed with subsequent medium changes, leaving preadipocytes as an adherent cell population.
2. At confluence, remove medium from cell culture dish. Rinse with 5 mL PBS and then remove PBS. Add 2.5 mL of trypsin solution and incubate for 5–10 min at 37 °C, periodically gently swirling and tapping the culture dish to detach cells. When 90 % of cells have detached, add 10 mL expansion medium to cell culture dish, and swirl dish to mix. Transfer the 12.5 mL cell suspension to a T150 flask and add an additional 12.5 mL of expansion medium to a total volume of 25 mL. Place in an incubator, change medium every 2–3 days until 90 % confluent, usually 2–3 days.
3. Remove medium from T150 flask. Rinse with 10 mL PBS and then remove PBS. Add 5 mL trypsin solution to flask and remove cells as above. When 90 % of cells have detached, add 35 mL of expansion medium to flask and swirl flask to mix. Divide the 50 mL cell suspension into eight new T150 flasks (5 mL per flask). Add an additional 20 mL of expansion medium to each flask to a total volume of 25 mL. Place in an

incubator, change medium every 3 days until 95 % confluent, usually 2–5 days. Time to confluence is variable and patient-specific.

4. Remove medium from flasks. Rinse each flask with 10 mL 1× PBS then remove PBS. Add 5 mL trypsin solution to each flask and remove cells as above. When 90 % of cells have detached, add 5 mL expansion medium to each flask and swirl to mix. Combine cell suspensions from all flasks into two 50 mL conical tubes and centrifuge for 10 min, 250 rcf, 4 °C. Remove medium, combine cell pellets by resuspending in 10 mL expansion medium. Count cells and repeat centrifugation step. Remove medium and resuspend cell pellet in freezing medium at 2×10^6 cells/mL.
5. Aliquot 1 mL of cell suspension into 2 mL screw-capped tubes. Transfer to −80° freezer for 24 hrs, then transfer tubes to liquid nitrogen storage unit for long-term storage (*see* **Note 2h**).

3.3 In Vitro Differentiation of Human Adipocytes

1. Thaw a frozen vial of PA containing two million cells by gently swirling cryovial in a 37 °C water bath; resuspend in 20 mL of expansion medium, in a single T150 flask, culture at 37 °C for 2–3 days until 90 % confluence.
2. Trypsinize, centrifuge, and resuspend PA in 10 mL of expansion medium; count cells using a hemocytometer and trypan blue exclusion. Seed cells at a density of 60,000 cells per well in a 24-well tissue culture plate (3×10^4 cells/cm^2) in 0.5 mL of expansion medium.
3. Remove expansion medium when cells are 90 % confluent usually at day 2–3; wash with 1× PBS and add differentiation medium. Replenish with fresh differentiation medium every 2–3 days.
4. Remove differentiation medium after 3–7 days; wash with 1× PBS and add maintenance medium (*see* **Note 3i**). Replenish with fresh maintenance medium every 2–3 days. Cells will begin to accumulate visible lipid droplets by day 10. By day 14, cells are lipid-laden and ready for experimental use (*see* **Notes 3j–l**, Fig. 4).

3.4 Oil Red-O Staining

1. Aspirate medium from cells that have been differentiated in a 24-well plate and wash once with 500 μL 1× PBS.
2. Add 200 μL 4 % formalin, fix for 15 min.
3. Aspirate formalin, and wash samples twice with 1× PBS
4. Add 200 μL of 60 % isopropanol, incubate for 5 min.
5. Aspirate 60 % isopropanol, and add 200 μL Oil Red-O working solution (*see* **Notes 4m** and **4n**).
6. Stain with Oil Red-O working solution for 15 min (*see* **Note 4o**).

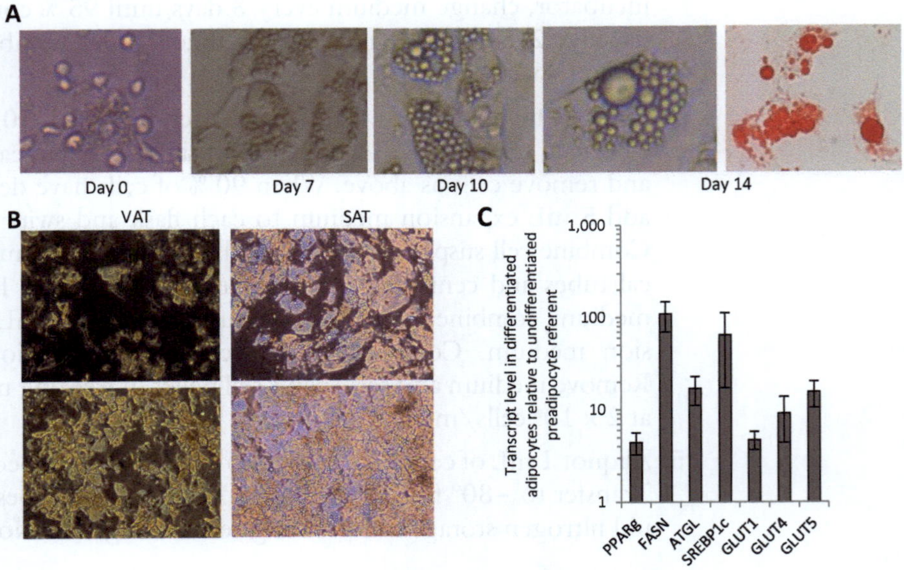

Fig. 4 Differentiated human adipocytes: (**a**) Human visceral adipocytes at various stages of differentiation; far right- Oil Red-O stained adipocytes. (**b**). Examples of VAT and SAT adipocytes from two different patients (*top*, *bottom*) at day 14 of differentiation; note greater degree of differentiation in SAT adipocytes, and moderate and good differentiation efficiencies (*top*, *bottom* respectively) between patients. (**c**). QRTPCR for adipogenic genes in RNA from VAT adipocytes after 14 days of differentiation; ordinate: fold difference in transcript levels in mature adipocytes compared with undifferentiated SVF referent; $n = 5$ samples

7. Aspirate Oil Red-O solution from wells (try to aspirate as much as possible in this step to avoid formation of precipitate).
8. Wash with 500 μL 1× PBS three times.
9. Image cells with light microscopy.
10. Aspirate PBS from well and allow to dry completely.
11. Add 200 μL 100 % isopropanol to well; place plate on rocker to disperse 60 % isopropanol for 15 min.
12. Aliquot 100 μL of solubilized Oil Red-O in isopropanol from each well in a well of a 96-well plate and read absorbance at 525 nM on a microplate spectrophotometer.

3.5 Lipolysis Assay

1. Culture cells as desired. For measurement of lipolysis, we seed 60,000 preadipocytes/well in a 24-well plate in 0.5 mL medium and differentiate into mature adipocytes as outlined above, then once cells are fully differentiated, proceed with lipolysis assay.
2. Remove maintenance medium and wash cells twice with warm 1× PBS; add 0.5 mL maintenance medium (without insulin) with or without 3 μM isoproterenol, then culture adipocytes for 6–72 h, then collect culture supernatants.

3. Pipet 2 μL of each supernatant into a 96-well plate. Reserve at least two wells to allow for one blank and one standard. The blank used is distilled H_2O; the glycerol standard solution is provided in the Triglyceride Determination Kit (*see* **Notes 4p–r**).
4. Add 270 μL of free glycerol reagent from the Triglyceride Determination Kit to each well, pipetting to mix. Incubate plate at 37 °C for 5 min.
5. Read plate at 540 nm on a microplate spectrophotometer.
6. Calculate the concentration of glycerol: (absorbance of sample − absorbance of blank) /(absorbance of standard − absorbance of blank) × 2.5 mg/mL = concentration of sample.

3.6 Labeled Acetate Lipogenesis Assay

1. Culture cells as desired. For measurement of basal lipogenesis, we seed 60,000 preadipocytes/well in a 24-well plate in 0.5 mL medium and differentiate into mature adipocytes as outlined above, then once cells are fully differentiated, culture for 72 h in maintenance medium and proceed with lipogenesis assay.
2. Remove maintenance medium and wash cells twice with warm 1× PBS; add 0.5 mL serum starvation medium supplemented with 100 nM insulin; incubate cells for 24 h at 37 °C.
3. Remove serum starvation medium; add 0.5 mL Lipogenesis medium; incubate at 37 °C overnight.
4. Wash cells twice with 1× PBS. Add 120 μL 0.1 N HCl to each well and pipette vigorously to lyse cells. Reserve 10 μL of lysate for protein assay (assess via Bradford assay); transfer 100 μL of remaining lysate to a 1.5 mL microcentrifuge tube.
5. Add 500 μL of 2:1 chloroform:methanol (v/v). Vortex briefly and incubate at room temperature for 5 min. Add 250 μL H_2O, vortex, and incubate at room temperature for an additional 5 min. Centrifuge samples for 10 min at 3000×*g*, 25 °C.
6. Carefully transfer lower lipid/organic phase to a scintillation vial containing 2 mL liquid scintillation fluid; measure ^3H activity on scintillation counter. Normalize cpm to protein concentration as determined by protein assay (*see* **Note 4s**).

3.7 Glucose Uptake Assay

1. Culture cells per desired experimental measurement in 24-well plates. For measurement of glucose uptake, we seed 60,000 preadipocytes/well in a 24-well plate in 0.5 mL medium and differentiate into mature adipocytes as outlined above, then once cells are fully differentiated, proceed with glucose uptake assay.
2. Remove medium and wash cells once with 1× PBS; add 0.5 mL/well serum starvation medium and incubate at 37 °C for 12 h.

3. Remove medium and wash cells twice with 1× PBS; add 0.5 mL/well 1 % BSA in PBS and incubate at 37 °C for 2 h.
4. Wash cells once with 1× PBS, add 0.5 mL/well 1× PBS with or without 100 nM insulin (exclude insulin for basal experimental arm), and incubate at 37 °C for 40 min.
5. Aspirate PBS, add 0.5 mL/well 1× PBS, 0.1 mM 2-deoxy glucose, 2 μCi/mL deoxy-D-glucose, 2- [1,2- ^{3}H(N)], with or without 200 nM insulin (exclude insulin for basal experimental arm), and incubate at 37 °C for 40 min.
6. Remove medium and wash cells three times with 1× PBS; add 420 μL 1 % SDS solution, lyse cells with vigorous pipetting; incubate at 25 °C for 10 min.
7. Collect 10 μL from each well for Bradford protein assay. Transfer 400 μL of whole cell lysate into 2 mL scintillation fluid in a scintillation vial. Count activity on scintillation counter.

4 Notes

4.1 Isolation of SVF from human adipose tissue

(a) Typical SVF yields from the above protocol are variable and patient-dependent, average yields are ~0.7 million cells per gram of tissue and range from 0.3 to 1.0 million per gram of tissue.

(b) Type I[13, 14] and type II[1, 5]collagenases, as well as liberase, a combination of type I and II collagenases [13], may be used for human SVF isolation. Vendors include Sigma Inc., Worthington Inc., Gibco Inc., and Roche Inc. (Liberase). In our experience, cell yields are incrementally higher with type II collagenase.

(c) The duration and concentration of collagenase digestion affect cell yield and viability. Yields increase and viability decreases with increasing digestion times and collagenase concentrations. We have experimented with 30-60 min digestion times and working (1×) collagenase concentrations of 0.5–3 mg/mL, ranges similar to other published protocols. In addition, collagenase activity varies among different lots, and some protocols calculate collagenase concentrations in units/mL rather than mg/mL. Type II collagenase (Sigma Inc.) typically ranges from 200 to 275 units/mg depending on lot. Digestion times, collagenase concentrations, and activity of specific lots of collagenase are variables each laboratory should consider during protocol optimization.

(d) Collagenase digestion may eliminate specific cell surface markers on leukocytes and preadipocytes. For SVF flow cytometry analysis, variable digestion times and collagenase concentrations should be tested to confirm that specific cell surface markers of interest are not eliminated by collagenase digestion.

(e) Particularly fibrous or less digested tissue may generate cell suspensions with excessive debris, elimination of which is helpful if SVF flow cytometry is planned (but not necessary for preadipocyte expansion, as debris is eliminated with medium changes during culture). Three interventions may aid in removing excessive debris from the final SVF suspension: first, an additional wash step may be added after RBC lysis; second, an additional filtration step may be added after mincing of tissue but before digestion; finally filtration of the SVF pellet retrieved after centrifugation may be performed [15]. Use of 250 μM, 150 μM, or 100 μM cell strainers is reported [13, 15, 16].

(f) Tissue may be stored overnight at 4 °C prior to processing, but with reduced cell yields and viability.

4.2 Expansion and storage of preadipocytes

(g) If starting adipose tissue amounts or SVF yields are low, then seed 50,000 SVF cells into a single well of a 6-well cell culture plate, then expand this into a 10 cm plate, then proceed with further expansion as outlined above into T150 flasks.

(h) Frozen preadipocytes may be stored for at least 6 months, beyond which viability decreases.

4.3 In vitro differentiation of human adipocytes

(i) Published protocols vary with respect to durations of culture in differentiation and maintenance media. Initial culture for 3–4 days in expansion medium containing serum is necessary to allow preadipocytes to adhere, after which differentiation is induced by changing cells to serum-free differentiation medium for 3–7 days or longer, after which culture is continued in serum-free maintenance medium to maintain ongoing differentiation. Differentiation and maintenance media contain similar base constituents, but the former contains IBMX and a thiazolidinedione. Some protocols include dexamethasone in differentiation and maintenance media [13], while we and others include dexamethasone only in differentiation medium. We observe better differentiation in some but not all samples with longer periods of culture in differentiation medium, and we typically culture cells for 7 days in differentiation medium, and up to 14 days if differentiation is poor. Others have reported similar longer periods of culture in differentiation medium [17, 18]. Recombinant human fibroblast growth factors (FGF-1, FGF-2) have been reported to enhance preadipocyte proliferation and adipocyte differentiation, but are not standard components of human adipocyte differentiation protocols as of this writing [19–24].

(j) Adipocytes are usually fully differentiated within 14 days, by which time they contain multiple lipid droplets that begin to coalesce into larger droplets. Longer periods of culture will result in further development of dominant and in some cases unilocular lipid droplets, but as time passes, adipocytes become less adherent and more difficult to manipulate in downstream assays. Care must be taken not to detach cells from plates during downstream assays, especially with longer periods of differentiation. We typically use adipocytes for assays after 14 days of differentiation, but cells can be used up to 4 weeks after induction of differentiation.

(k) We assess differentiation using Oil Red-O staining, QRTPCR for adipogenic gene expression, and lipolysis assay. We quantify lipogenesis using both quantification of Oil Red-O staining and incorporation of ^{3}H–acetate into lipid. Mature adipocytes accumulate Oil-Red-O in lipid droplets, markedly increase adipogenic gene expression, and demonstrate lipolytic capacity manifested by glycerol release in response to beta-adrenergic stimulation (Fig. 4).

(l) Significant patient-specific variability is observed in degree of differentiation as assessed by the percentage of preadipocytes that accumulate cytoplasmic lipid. Depot-specific differences in differentiation efficiency are also observed, typically 50–70 % for VAT and 90 + % for SAT.

4.4 Metabolic phenotyping

(m) Oil Red-O solution will precipitate/adhere to the filter; prepare double what you intend to use that day to account for this. Stain immediately after preparation/filtration to avoid precipitation of Oil Red-O.

(n) For Oil Red-O staining, dispense all reagents with a pipette against the side of the tissue culture well, as pipetting directly onto cells will disrupt the cell monolayer.

(o) Oil Red-O staining longer than 15 min should be avoided, as this will lead to precipitation of Oil Red-O and falsely elevated readings.

(p) The Triglyceride Determination Kit for lipolysis assay measures free glycerol via a coupled enzymatic reaction that produces a quinone imine dye, absorbance of which at 540 nm is directly proportional to glycerol concentration.

(q) Free glycerol reagent for lipolysis assay comes pre-aliquoted in powdered form and should be prepared for use by adding 40 mL of H_2O to one bottle of powdered reagent. Gently invert the bottle to mix the reagent. Each bottle can be stored at 4 °C for up to 60 days after it is reconstituted. This reagent should be colorless; discard bottle if a purple tint is observed.

Free glycerol reagent should not be vigorously shaken or vortexed, as this will deactivate it.

(r) Place the standard in the lipolysis assay at least one well away from other samples on the 96-well plate, as its strong signal may affect absorbance readings of other samples.

(s) Lipogenesis may be quantified by measurement of incorporation of labeled fatty acid precursors into cellular lipids. Many assays utilize C^{14} labeling, but we have found good results with the above tritium-labeled acetate-based assay, which is modified from previously published protocols used in adipocytes, hepatocytes, and other cell types [25, 26]. Incorporation into lipid involves transfer of a—CH2 moiety, and while some label is lost by H-D exchange, we have found the assay to be reproducible and reliably distinguish patient-specific adipocyte lipogenic capacities.

Funding

Supported by NIH grants R01DK097449 (R.W.O.), DK090262 (C.N.L.), T32DK101357, F32DK105676 (L.A.M.).

References

1. O'Rourke RW, Meyer KA, Gaston G, White AE, Lumeng CN, Marks DL (2013) Hexosamine biosynthesis is a possible mechanism underlying hypoxia's effects on lipid metabolism in human adipocytes. PLoS One 8:e71165
2. Tchkonia T, Giorgadze N, Pirtskhalava T, Thomou T, DePonte M, Koo A, Forse RA, Chinnappan D, Martin-Ruiz C, von Zglinicki T, Kirkland JL (2006) Fat depot-specific characteristics are retained in strains derived from single human preadipocytes. Diabetes 55:2571–2578
3. Tchkonia T, Tchoukalova YD, Giorgadze N, Pirtskhalava T, Karagiannides I, Forse RA, Koo A, Stevenson M, Chinnappan D, Cartwright A, Jensen MD, Kirkland JL (2005) Abundance of two human preadipocyte subtypes with distinct capacities for replication, adipogenesis, and apoptosis varies among fat depots. Am J Physiol Endocrinol Metab 288:E267–E277
4. Tchkonia T, Giorgadze N, Pirtskhalava T, Tchoukalova Y, Karagiannides I, Forse RA, DePonte M, Stevenson M, Guo W, Han J, Waloga G, Lash TL, Jensen MD, Kirkland JL (2002) Fat depot origin affects adipogenesis in primary cultured and cloned human preadipocytes. Am J Physiol Regul Integr Comp Physiol 282:R1286–R1296
5. Tchoukalova YD, Koutsari C, Votruba SB, Tchkonia T, Giorgadze N, Thomou T, Kirkland JL, Jensen MD (2010) Sex- and depot-dependent differences in adipogenesis in normal-weight humans. Obesity (Silver Spring) 18:1875–1880
6. Lee YH, Petkova AP, Granneman JG (2013) Identification of an adipogenic niche for adipose tissue remodeling and restoration. Cell Metab 18:355–367
7. Sengenès C, Lolmède K, Zakaroff-Girard A, Busse R, Bouloumié A (2005) Preadipocytes in the human subcutaneous adipose tissue display distinct features from the adult mesenchymal and hematopoietic stem cells. J Cell Physiol 205:114–122
8. Wu J, Boström P, Sparks LM, Ye L, Choi JH, Giang AH, Khandekar M, Virtanen KA, Nuutila P, Schaart G, Huang K, Tu H, van Marken Lichtenbelt WD, Hoeks J, Enerbäck S, Schrauwen P, Spiegelman BM (2012) Beige adipocytes are a distinct type of thermogenic fat cell in mouse and human. Cell 150: 366–376
9. Virtanen KA, Lidell ME, Orava J, Heglind M, Westergren R, Niemi T et al (2009) Functional brown adipose tissue in healthy adults. N Engl J Med 360:1518–1525

10. Carey VJ, Walters EE, Colditz GA et al (1997) Body fat distribution and risk of non-insulin-dependent diabetes mellitus in women. Nurses' Health Study. Am J Epidemiol 145:614–619
11. Kissebah AH, Vydelingum N, Murray R, Evans DJ, Hartz AJ, Kalkhoff RK et al (1982) Relation of body fat distribution to metabolic complications of obesity. J Clin Endocrinol Metab 54:254–260
12. Tran TT, Yamamoto Y, Gesta S, Kahn CR (2008) Beneficial effects of subcutaneous fat transplantation on metabolism. Cell Metab 7:410–420
13. Lee MJ, Fried SK (2014) Optimal protocol for the differentiation and metabolic analysis of human adipose stromal cells. Methods Enzymol 538:49–65
14. Lystedt E, Westergren H, Brynhildsen J, Lindh-Astrand L, Gustavsson J, Nystrom FH et al (2005) Subcutaneous adipocytes from obese hyperinsulinemic women with polycystic ovary syndrome exhibit normal insulin sensitivity but reduced maximal insulin responsiveness. Eur J Endocrinol 153:831–835
15. Hutley LJ, Herington AC, Shurety W, Cheung C, Vesey DA, Cameron DP, Prins JB (2001) Human adipose tissue endothelial cells promote preadipocyte proliferation. Am J Physiol Endocrinol Metab 281:E1037–E1044
16. Hauner H, Röhrig K, Petruschke T (1995) Effects of epidermal growth factor (EGF), platelet-derived growth factor (PDGF) and fibroblast growth factor (FGF) on humanadipocyte development and function. Eur J Clin Invest 25(2):90–96
17. Dicker A, Ryden M, Naslund E, Muehlen IE, Wiren M, Lafontan M et al (2004) Effect of testosterone on lipolysis in human pre-adipocytes from different fat depots. Diabetologia 47:420–428
18. Yu G, ZE F, Wu X, Hebert T, Halvorsen YD, Buehrer BM et al (2011) Adipogenic differentiation of adipose-derived stem cells. Methods Mol Biol 702:193–200
19. Hutley LJ, Shurety W, Newell F, McGeary R, Pelton N, Grant J, Herington A, Cameron D, Whitehead J, Prins J (2004) Fibroblast growth factor 1: a key regulator of human adipogenesis. Diabetes 53:3097–3106
20. Kakudo N, Shimotsuma A, Kusumoto K (2007) Fibroblast growth factor-2 stimulates adipogenic differentiation of human adipose-derived stem cells. Biochem Biophys Res Commun 359:239–244
21. Newell FS, Su H, Tornqvist H, Whitehead JP, Prins JB, Hutley LJ (2006) Characterization of the transcriptional and functional effects of fibroblast growth factor-1 on human preadipocyte differentiation. FASEB J 20:2615–2617
22. Skurk T, Ecklebe S, Hauner H (2007) A novel technique to propagate primary human preadipocytes without loss of differentiation capacity. Obesity (Silver Spring) 15:2925–2931
23. Skurk T, Hauner H (2012) Primary culture of human adipocyte precursor cells: Expansion and differentiation. Methods Mol Biol 806:215–226
24. Widberg CH, Newell FS, Bachmann AW, Ramnoruth SN, Spelta MC, Whitehead JP, Hutley LJ, Prins JB (2009) Fibroblast growth factor receptor 1 is a key regulator of early adipogenic events in human preadipocytes. Am J Physiol Endocrinol Metab 296:E121–E131
25. Akie TE, Cooper MP (2015) Determination of fatty acid oxidation and lipogenesis in mouse primary hepatocytes. J Vis Exp 102:e52982
26. Perez-Diaz S, Johnson LA, DeKroon RM, Moreno-Navarrete JM, Alzate O, Fernandez-Real JM, Maeda N, Arbones-Mainar JM (2014) Polymerase I and transcript release factor (PTRF) regulates adipocyte differentiation and determines adipose tissue expandability. FASEB J 28:3769–3779

Chapter 8

Protocols for Generation of Immortalized Human Brown and White Preadipocyte Cell Lines

Farnaz Shamsi and Yu-Hua Tseng

Abstract

Human brown and white preadipocytes offer unique cell models to study human adipogenesis and thermogenesis. Here, we describe the detailed procedures for isolation of human brown and white predipocytes from deep and superficial neck fat. To grow these cells in vitro for a prolonged period of time, they should be immortalized following the procedure discussed here. We also provide the protocol for expansion, cryopreservation, and adipogenic differentiation of cells.

Key words Brown and white adipose tissue, Stromal vascular fraction (SVF), Preadipocytes, Neck fat, Immortalization, Differentiation

1 Introduction

Human obesity results from an imbalance between energy intake and expenditure, and it is a major contributor to metabolic syndrome and disorders, such as type 2 diabetes, cardiovascular disease, and some cancers. Brown and white adipose tissues are two functionally distinct types of fat present in mammals. While the main role of white adipose tissue (WAT) is storing excess calories, brown adipose tissue (BAT) specializes in energy dissipation and thermogenesis through the activity of uncoupling protein 1 (UCP1). BAT can be activated by cold stimulation in both humans and rodents. Cold-activated BAT in humans consumes more glucose per tissue weight than any other tissue [1]. Given the ability of BAT to utilize fuel and transform chemical energy of nutrients into heat, increasing the activity or amount of human brown fat constitutes a promising strategy to combat obesity.

The presence of functional BAT in adult humans was recently rediscovered by using 18F-fluorodeoxyglucose positron emission tomography/computed tomography (18F-FDG PET/CT) as bilaterally symmetric patches of radio-labeled glucose uptake in the neck and supraclavicular region [2]. Later, UCP1-expressing

Jun Wu (ed.), *Thermogenic Fat: Methods and Protocols*, Methods in Molecular Biology, vol. 1566,
DOI 10.1007/978-1-4939-6820-6_8, © Springer Science+Business Media LLC 2017

adipocytes were reported to exist in adult humans around the neck, supraclavicular, and spinal cord regions [3–5]. Most importantly, the amount of brown adipose tissue was shown to be inversely correlated with body-mass index and percentage of body fat, whereas positively correlated with resting metabolic rate in human subjects.

Examination of adipose tissue collected from different depots of the human neck revealed that superficial and deeper fat had the classical histological, ultrastructural, and gene expression features of rodent white and brown adipose tissue, respectively. Importantly, deeper human neck fat expressed higher levels of UCP1 compared to the more superficial neck depots [6]. To enable comprehensive studies of molecular and cellular aspects of human adipogenesis and thermogenesis in vitro, we have established pairs of immortalized human brown and white preadipocytes from human neck fat biopsies. Procedures of generation, growth, and differentiation of these cells are detailed below.

2 Materials

2.1 Equipment

1. Laminar flow hood.
2. Humidified CO_2 incubator.
3. Water bath.
4. Centrifuge.
5. pH meter.
6. Chemical balance.
7. Stirrer.
8. Cell Culture vessels.
9. Serological pipettes.
10. Aspiration pipettes.
11. Inverted microscope.
12. "Mr. Frosty" freezing container.
13. Liquid Nitrogen tank.
14. Hemocytometer or automated cell counter.
15. 100 μm Cell Strainer.
16. Parafilm.
17. Sterile dissection instruments.

2.2 Reagents

1. Digestion solution: 2 mg/mL Collagenase-I in 3.55 BSA in PBS. Dissolve 20 mg Collagenase-I in 10 mL PBS buffer. Add BSA to final concentration of 3.5 % to the solution. Filter using a 0.22 μM filter unit. Prepare approximately 10 mL per sample.

2. Growth medium: 10 % FBS in DMEM/High glucose. Add 100 mL Fetal Bovine Serum (FBS) to 890 mL DMEM/High Glucose. Add 10 mL Penicillin-Streptomycin. pH to 7.4 using either HCl or NaOH and sterile filter. For preparation of growth medium with gentamicin, add 1000× stock solution to growth medium and sterile filter.
3. Freezing medium: 50 % FBS and 10 % DMSO in DMEM/High glucose. Add 25 mL FBS to 20 mL DMEM/High Glucose. Add 5 mL DMSO. Sterile filter.
4. Differentiation and Induction medium: Prepare the reagent stocks for differentiation and induction medium as follow: Biotin: Dissolve 0.08 g in 10 mL 0.1 N NaOH to make 30 mM solution. Insulin: Dissolve Insluin in cncentration of 10 mg/mL in 0.01 MHCl. Pantothenate: Dissolve 8.5 mM solution in H_2O (500× stock). Dexamethasone: Dissolve 0.002 g in 1 mL EtOH (2 mg/mL; 5 mM). Isobutyl methylxanthine (IBMX): Make 50 mM stock in 0.1 M KOH. 3,3′,5-Triiodo-L-thyronine (T_3): Add 1.0 mL of 1.0 N NaOH to 1 mg, gently swirl to dissolve the powder and add 152.6 mL of sterile medium (10 μM). Indomethacin: Dissolve 447.25 mg Indomethacin in 10 mL EtOH (125 mM). Heat upto 75 °C. Aliquot stock solutions and store them at −20 °C. Avoid multiple freeze/thaw cycles. Differentiation Medium: Add 1 mL FBS to 48.5 mL DMEM/High Glucose. Add 500 μl Penicillin-Streptomycin, 15 μl human Insulin (0.5 μM), and 10 μl T3 (2 nM). pH to 7.4 using either HCl or NaOH and sterile filter. Induction Medium: Add 1 mL FBS to 48.5 mL DMEM/High Glucose. Add 500 μl Penicillin-Streptomycin, 50 μl Biotin (33 μM), 15 μl human Insulin (0.5 μM), 100 μl Pantothenate (17 μM), 1 μl Dexamethasone (0.1 μM), 10 μl T3 (2 nM), 500 μl IBMX (500 μM), and 12.5 μl Indomethacin (30 μM). pH to 7.4 using either HCl or NaOH and sterile filter.
5. Trypsin–EDTA (0.25 %).
6. PolyJet In Vitro DNA Transfection Reagent.
7. pBabe-hTert-hygro or pBabe-hTert-neo retroviral plasmids (Addgene, Cambridge, MA).

3 Methods

3.1 Isolation of Brown and White Fat Progenitors from Human Neck Fat.

All the steps should be carried out under a laminar flow hood.

1. Remove fat sample from collection tube and transfer onto a piece of parafilm.
2. Add a small amount of digestion solution to the tissue to avoid drying.

3. Mince the tissue in the solution to a fine consistency.
4. Transfer to a 50 mL tube containing 10 mL digestion solution and seal the tube with parafilm. For isolation of human BAT progenitors (hBAT-SVF cells), tissues collected from deep neck fat (carotid sheath, longus colli, and prevertebral depots) should be combined. For isolation of human WAT progenitors (hWAT-SVF cells), tissues collected from superficial neck fat (subcutaneous and subplatysmal depots) should be combined.
5. Vortex each tube thoroughly and put the tubes in the water bath to shake at 70 rpm and 37 °C. Make sure that the water bath is full enough so that the tissue is submerged.
6. Vortex the tubes thoroughly every 10 min. The tubes should remain in the bath until all lumps appear to be dissolved, but not so long that a clear fat supernatant layer appears (up to 45 min—longer digests tend to result in more cells, but over-digestion can decrease the cell yield).
7. When digestion is complete, add 3 mL FBS to stop Collagenase and pipet thoroughly.
8. Pass through a 100 μm filter into a fresh 50 mL tube (wash the tube again with 10 mL growth medium, then filter).
9. Centrifuge the tubes at 300 × *g* for 10 min.
10. Gently aspirate off most of the supernatant and resuspend the pellet in 10 mL growth medium.
11. Centrifuge at 300 × *g* for 10 min.
12. Gently aspirate off the most of the supernatant leaving only the pellet.
13. Resuspend the pellet in 10 mL growth medium with gentamicin.
14. Centrifuge at 300 × *g* for 10 min.
15. Gently aspirate off most of the supernatant and repeat wash/spinning again.
16. Gently aspirate off most of the supernatant.
17. Resuspend the pellet in 1 mL growth medium with gentamicin.
18. Seed the cells in growth medium in one well of a 12-well plate. Add 1 mL growth medium to wash the tube again, then transfer to the same well.
19. The next day, carefully wash the cells only one time with growth medium + gentamicin (some cells might be loosely attached to the plate, make sure not to detach them while washing).
20. 48 h later, repeat washing.
21. Remove Gentamicin from the growth medium after 72 h.

3.2 Generation of Immortalized Human Brown and White Fat Progenitors

In this protocol, a retroviral plasmid (pBabe-hTert-hygro or pBabe-hTert-neo) is used to deliver human Telomerase reverse transcriptase (hTERT) to primary hBAT and hWAT-SVF cells. For production of high titer retrovirus for infection, helper-free Amphotropic Phoenix (Phoenix-A) cell line can be used. Phoenix Ampho and Eco packaging cell lines are second-generation retrovirus producer lines for the generation of helper-free amphotropic and ecotropic retroviruses. Phoenix-A Cell Line can be used to deliver genes to dividing cells of most mammalian species, including human (*see* **Note 1**) [7, 8].

Day 0: Preparation of Phoenix-A Retrovirus Producer cells for Transfection

1. 18–24 h prior to transfection, seed Phoenix-A cells at 4–5 million cells per 10-cm plate in growth medium.
2. After plating cells, transfer plates to the incubator. Do not disturb the cells for several hours to ensure their attachment.

Day 1: Transfection

3. 1 h prior to transfection, change the medium of Phoenix-A cells and add 6 mL fresh growth medium to each 10-cm plate. Be careful not to detach the cells.
4. For each 10-cm plate, dilute 5 μg of DNA into 250 μl of serum-free DMEM with High Glucose. Gently pipette up and down or vortex briefly to mix.
5. For each 10-cm plate, dilute 15 μl of PolyJet™ reagent into 250 μl of serum-free DMEM with High Glucose. Pipette up and down 3–4 times to mix.
6. Immediately add the diluted PolyJet™ reagent solution to the diluted DNA. Pipette up and down 3–4 times or vortex briefly to mix. Let the mixture stand for 10–15 min at room temperature to allow DNA/reagent complexes to form.
7. Add the 500 μl DNA/reagent mixture to the plate in a drop-wise manner and homogenize the mixture by gently swirling the plate.
8. 6–12 h post-infection, remove medium containing DNA/reagent mixture and add fresh growth medium to the cells.

Day 2: Preparation of primary hBAT-SVF and hWAT-SVF cells for infection

9. Plate 4–8 × 10^4 cells in one well of a 12-well plate. To achieve maximum infection efficiency, cells should be around 60–80 % confluent. Make sure to include one control well for drug selection.

Day 3: Collection of retrovirus particles and infection of primary hBAT-SVF and hWAT-SVF cells

10. Harvest the Phoenix-A cells supernatant containing retrovirus particles.

11. Filter through 0.45 μm filter to removes cells in the suspension.
12. Virus can be used freshly for infecting the target cells. Alternatively, aliquot the virus and immediately freeze at −80 °C for later infection. Avoid freezing and thawing of virus as it decreases virus titer (*see* **Note 2**).
13. For infection of hBAT-SVF and hWAT-SVF cells, dilute pBabe-hTert-hygro or pBabe-hTert-neo virus in growth medium in 1:1 ratio. Add polybrene to a final concentration of 4 μg/mL. Add the mixture to cells.
14. Repeat **step 13** every day while the cells are growing until the cells reach near 90 % confluency (usually three times).
15. Once cells are around 90 % confluent, split the cells in each well into two wells.
16. The next day, start drug selection with the proper selectable marker. If using hygromycin, start with 100 μg/mL hygromycin in growth medium. For Neomycin selection, start with 500 μg/mL G418 (Geneticin). The optimal concentration of antibiotic can vary between subjects, and should be determined for each cell line (*see* **Note 3**).
17. Closely monitor cells' response to the selection drug. A large number of control cells should die within 2–3 days. If little or no cell death is observed, increase drug concentration. To ensure effective selection, change medium containing drug every other day.
18. Once all of the control cells (no infection) die, remove the antibiotic and let the cells grow without antibiotic for several days. Immortalized human preadipocytes will start to proliferate and clusters of cells will appear after almost a week.
19. When cells reach 80-90% confluency, split the cells or freeze them according to the freezing protocol described below.

3.3 Maintenance and Expansion of Immortalized Human Preadipocytes

1. Grow the cells in growth medium at 37 °C in a humidified incubator with 5 % CO_2.
2. When cells reach 80–90 % confluency, split them 1:2 or 1:3 (*see* **Notes 4** and **5**).
3. To subculture, aspirate culture medium. Rinse once with Calcium and Magnesium free PBS. For a 10 cm dish, add 1 mL of 0.25 % (w/v) pre-warmed trypsin and return to incubator for 2–3 min. Observe cells under an inverted microscope. When cells start to become round and detach from the plate, neutralize trypsin by adding 10 mL of growth medium to the dish and collect cells by gentle pipetting. Plate the cells in new dishes and place them in the incubator to adhere.
4. Change culture media the following day to remove floating cells.

3.4 Cryopreservation of Primary and Immortalized Human Preadipocytes

1. Prepare freezing medium as described above.
2. Gently detach cells from the tissue culture vessel following the procedure described above for subculture.
3. After adding growth media and collecting cells by pipetting, transfer cell solution to a 15 mL tube and centrifuge at 300 × *g* for 4 min.
4. Gently remove the tube from centrifuge, aspirate most of the supernatant without disturbing the cell pellet, and resuspend pellet in 1 mL freezing media per vial to be frozen. To ensure maximum recovery of frozen cells, each confluent 10-cm dish should be frozen in two vials.
5. Place vials in "Mr. Frosty" freezing container containing 100 % isopropyl alcohol at −80 °C for 24 h.
6. Transfer vials to liquid N_2 tank for indefinite storage.

3.5 Differentiation of Immortalized Human White and Brown Preadipocytes

Seed 40,000–50,000 cells per well in a 24-well plate.

1. Let the cells grow in growth medium until they are fully confluent (usually 3–4 days for most cell lines, up to 6 days) (*see* **Note 6**).
2. Once cells reach confluency, aspirate growth media and add freshly prepared induction medium to cells.
3. Add fresh induction media every 3 days for 12–18 days.
4. Harvest differentiated adipocytes for further analysis (Gene expression analysis, Oil red O staining, etc.).
5. For some lines, Rosiglitazone (1 μM) should be added in the induction medium to enhance differentiation.

Alternatively, to induce increased *UCP1* expression, mitochondrial activity, and fuel utilization in mature adipocytes, cells can be pretreated with BMP7 in differentiation medium prior to adipogenic differentiation as described below [9–11].

1. Add 3.3 nM BMP7 or similar volume of vehicle to the differentiation medium.
2. Once cells reach confluency, aspirate growth media, and add differentiation medium containing BMP7 or vehicle to cells.
3. Treat cells with BMP7 in differentiation medium for 6 days. Add fresh differentiation medium every 3 days.
4. At day 7, start adding induction media to cells.
5. Change induction medium every 3 days for 12–18 days.

4 Notes

1. Handling Phoenix cells: Never let Phoenix-A cells reach confluence, as it decreases their transfection efficiency. For maximally healthy culture condition, passage cells when they are 70–80 % confluent in 1:4 or 1:5 ratios.
2. Virus concentration: To increase the titer of retrovirus preparations, supernatant can be concentrated using commercially available retrovirus concentration kits or other in-house methods. Try to use fresh virus when possible, as the titer decreases with each freeze-thaw cycle.
3. Drug selection: Cells derived from different human subjects vary in their response to G418 or Hygromycin. Try different concentrations for each cell line. Monitor the infected and noninfected control cells closely while they are under drug selection.
4. Maintenance of immortalized human white and brown preadipocytes: To ensure an optimal growth condition, never plate the cells too sparse. Some cell lines tend to grow very slowly when they are below 20–30 % confluent. Splitting 80–90 % confluent culture into 1:2 to 1:3 ratios usually provides a good starting confluency.
5. Mycoplasma contamination: Test the cells for mycoplasma contamination on a regular basis. Stick to good aseptic technique to prevent contamination. Treat contaminated cells with an antimycoplasma reagent such as Plasmocin™ (InvivoGen, San Diego, CA) according to manufacturer's instruction.
6. Differentiation of immortalized human white and brown preadipocytes: Do not start adding the induction medium to cells before they are completely confluent. Sub-confluent preadipocytes do not differentiate efficiently. The optimal starting cell number can vary slightly between cell lines and it should be determined for each individual cell line.

Acknowledgments

This work was supported in part by NIH grants R01 DK077097 and R01 DK102898 (to Y.-H.T.), and P30 DK036836 (to Joslin Diabetes Research Center).

References

1. Orava J, Nuutila P, Lidell ME, Oikonen V, Noponen T, Viljanen T, Scheinin M, Taittonen M, Niemi T, Enerback S, Virtanen KA (2011) Different metabolic responses of human brown adipose tissue to activation by cold and insulin. Cell Metab 14(2):272–279. doi:10.1016/j.cmet.2011.06.012
2. Nedergaard J, Bengtsson T, Cannon B (2007) Unexpected evidence for active brown adipose tissue in adult humans. Am J Physiol Endocrinol Metab 293(2):E444–E452. doi:10.1152/ajpendo.00691.2006
3. Cypess AM, Lehman S, Williams G, Tal I, Rodman D, Goldfine AB, Kuo FC, Palmer EL, Tseng YH, Doria A, Kolodny GM, Kahn CR (2009) Identification and importance of brown adipose tissue in adult humans. N Engl J Med 360(15):1509–1517. doi:10.1056/NEJMoa0810780
4. van Marken Lichtenbelt WD, Vanhommerig JW, Smulders NM, Drossaerts JM, Kemerink GJ, Bouvy ND, Schrauwen P, Teule GJ (2009) Cold-activated brown adipose tissue in healthy men. N Engl J Med 360(15):1500–1508. doi:10.1056/NEJMoa0808718
5. Virtanen KA, Lidell ME, Orava J, Heglind M, Westergren R, Niemi T, Taittonen M, Laine J, Savisto NJ, Enerback S, Nuutila P (2009) Functional brown adipose tissue in healthy adults. N Engl J Med 360(15):1518–1525. doi:10.1056/NEJMoa0808949
6. Cypess AM, White AP, Vernochet C, Schulz TJ, Xue R, Sass CA, Huang TL, Roberts-Toler C, Weiner LS, Sze C, Chacko AT, Deschamps LN, Herder LM, Truchan N, Glasgow AL, Holman AR, Gavrila A, Hasselgren PO, Mori MA, Molla M, Tseng YH (2013) Anatomical localization, gene expression profiling and functional characterization of adult human neck brown fat. Nat Med 19(5):635–639. doi:10.1038/nm.3112
7. Pear WS, Nolan GP, Scott ML, Baltimore D (1993) Production of high-titer helper-free retroviruses by transient transfection. Proc Natl Acad Sci U S A 90(18):8392–8396
8. Swift S, Lorens J, Achacoso P, Nolan GP (2001) Rapid production of retroviruses for efficient gene delivery to mammalian cells using 293T cell-based systems. Curr Protoc Immunol Chapter 10:Unit 10 17C. doi:10.1002/0471142735.im1017cs31
9. Schulz TJ, Huang TL, Tran TT, Zhang H, Townsend KL, Shadrach JL, Cerletti M, McDougall LE, Giorgadze N, Tchkonia T, Schrier D, Falb D, Kirkland JL, Wagers AJ, Tseng YH (2011) Identification of inducible brown adipocyte progenitors residing in skeletal muscle and white fat. Proc Natl Acad Sci U S A 108(1):143–148. doi:10.1073/pnas.1010929108
10. Tseng YH, Kokkotou E, Schulz TJ, Huang TL, Winnay JN, Taniguchi CM, Tran TT, Suzuki R, Espinoza DO, Yamamoto Y, Ahrens MJ, Dudley AT, Norris AW, Kulkarni RN, Kahn CR (2008) New role of bone morphogenetic protein 7 in brown adipogenesis and energy expenditure. Nature 454(7207):1000–1004. doi:10.1038/nature07221
11. Xue R, Lynes MD, Dreyfuss JM, Shamsi F, Schulz TJ, Zhang H, Huang TL, Townsend KL, Li Y, Takahashi H, Weiner LS, White AP, Lynes MS, Rubin LL, Goodyear LJ, Cypess AM, Tseng YH (2015) Clonal analyses and gene profiling identify genetic biomarkers of the thermogenic potential of human brown and white preadipocytes. Nat Med 21(7):760–768. doi:10.1038/nm.3881

Part II

Functional Analysis, Bioengineering, and Other Applications

Chapter 9

Gene Expression and Histological Analysis of Activated Brown Adipocytes in Adipose Tissue

Yun-Hee Lee

Abstract

With the rediscovery of brown adipose tissue in adult humans, identification and characterization of brown adipocytes have been topics of great interest in the field of adipose tissue research. In particular, identification of the molecular mechanisms that activate thermogenic adipocytes suggests promising targets for increasing energy expenditure and ultimately combatting obesity and obesity-related metabolic disease. Thus, the methodology for identifying brown adipocytes in vivo is important for the precise determination of the metabolic activity of brown adipose tissue and de novo brown adipogenesis in white adipose tissue. In addition, in vivo analysis of brown adipocytes in combination with lineage tracing is essential to investigate the cellular origins of brown adipocytes. This chapter first provides a brief overview of lineage tracing studies performed in the search for the cellular origins of brown adipocytes. The chapter then describes the immunohistochemistry methodology for identifying brown adipocytes in adipose tissue, including analyses in histologic tissue sections and whole mount tissue. Lastly, it discusses flow cytometric analysis of dissociated cells from adipose tissue, and isolation of live adipocytes for subsequent gene expression profiling using fluorescence-activated cell sorting.

Key words Brown adipocyte, Adipose tissue, Lineage tracing, Immunohistochemistry, Flow cytometry

1 Introduction

Activation of thermogenesis in brown adipose tissue has been investigated as a potential target mechanism for increasing energy expenditure [1, 2]. In particular, rediscovery of human brown adipose tissue [2, 3] has led to increased research interest in the identity of brown adipocytes and the molecular mechanisms activating their thermogenesis. Thus, development of methodology for identifying brown adipocytes in vivo is important for the precise determination of the metabolic activity of brown adipocytes and brown adipose tissue. In addition to brown adipocytes, a new cell type has been identified, so-called beige adipocytes [4], which can be generated by de novo brown adipogenesis from progenitors in white adipose tissue [5–7]. Recognition of the complex phenotypes of

Jun Wu (ed.), *Thermogenic Fat: Methods and Protocols*, Methods in Molecular Biology, vol. 1566,
DOI 10.1007/978-1-4939-6820-6_9, © Springer Science+Business Media LLC 2017

brown/beige adipocytes appearing in various anatomical locations raises fundamental questions on their cellular origins.

Genetic labeling of progenitors and lineage tracing have been key methods for identifying the cellular origins of brown adipocytes. One of the most popular and powerful lineage tracing systems is based on the Cre-Lox system, consisting of a Cre-driver and a Cre-responder mouse line [8]. In the Cre-driver line, expression of Cre recombinase is under the control of a promoter of an active gene specifically expressed in progenitors. Thus, the identification of a specific gene that is highly expressed in progenitors is a prerequisite for establishing this system. In this regard, Lee et al. [5] reported platelet-derived growth factor receptor-alpha (PDGFRα) expression in bipotent brown and white adipocyte progenitors, and this observation led to the establishment of a model system for adipocyte progenitor tracing [9–11]. For lineage tracing, a Cre-driver line can be crossed with Cre-responder reporter lines (e.g., a knock-in of a floxed stop codon at the Rosa26-locus followed by fluorescent protein reporter expression) [8]. In this system, cell type specific Cre-recombinase expression results in permanent fluorescent tagging of the specific cell types and their progenies. Furthermore, by using drug-controllable Cre recombinase (e.g., CreERT2), genetic tagging can be used to label cells during a specific developmental time frame [8]. In addition to adipocyte progenitor tracing, genetic tracing of mature adipocytes using a Cre expressing line under the control of the adiponectin or uncoupling protein 1 (UCP1) promoter provided information on the dynamic behaviors of the convertible phenotypes of white and brown adipocytes [10, 12]. Membrane-targeted reporter proteins have been useful for genetic tracing of adipocytes, since the cellular boundary can be distinguished in histologic sections [9]. In combination with lineage tracing, fluorescence-activated cell sorting (FACS) technology has aided accurate gene expression profiling of in vivo brown adipocyte lineages by enabling isolation of fluorescently tagged subpopulations and analysis of a specific cell type at the single-cell level [13].

This chapter briefly introduces methods for activating brown adipocytes in vivo and describes examples of lineage tracing studies using chemical and genetic tagging in search of the cellular origins of brown adipocytes. The chapter describes the immunostaining methodology used to identify brown adipocytes in adipose tissue, including analyses in histologic tissue sections (e.g., paraffin sections, cryosections) and whole mount tissues. Finally, the chapter explains flow cytometric analysis of the dissociated adipocytes from adipose tissue composed of a mixed population of adipocytes and stromal vascular cells. In combination with live cell staining and adipocyte marker staining, FACS isolation of adipocytes and subsequent gene expression profiling are discussed as reliable methods for characterizing isolated adipocyte phenotypes.

2 Materials

2.1 Lineage Tracing

1. Pdgfra-CreERT2[14]/ROSA26-tdTomato (B6.Cg-*Gt(ROSA)26Sor tm14(CAG-tdTomato)Hze/*J) double transgenic mice.
2. Adipoq-CreERT2[10]/ROSA26-mTmG (B6.129(Cg)-*Gt(ROSA)26Sor tm4(ACTB-tdTomato,-EGFP)Luo/*J) double transgenic mice.
3. CL316,243 (CL): 1.5 mM solution in saline.
4. 5-Bromo-2′-deoxyuridine (BrdU): 80 mg/mL in 50 % DMSO.
5. Tamoxifen: 20 mg/mL in sunflower oil. To dissolve tamoxifen in sunflower oil, incubate the solution at 55 °C for 30 min with vortexing.
6. Mini-Osmotic Pumps (0.5 μL/h infusion rate).

2.2 Immunohistochemistry

1. Fixation buffer: 10 % neutral buffered formalin, 4 % paraformaldehyde in PBS, 20 % sucrose solution.
2. Tissue Processor: LEICA TP 1020.
3. Tissue-tek cryo OCT compound and Paraffin.
4. Embedding system: Sakura Tissue-Tek 5 embedding console system for paraffin blocks.
5. Microtome.
6. Solution for deparaffinization/rehydration of paraffin tissue sections.
   (a) 100 % xylene.
   (b) xylene:ethanol (1:1) mixture.
   (c) 100 % ethanol.
   (d) 95 % ethanol.
   (e) 70 % ethanol.
7. Antigen retrieval buffer: 10 mM sodium citrate in water, 0.05 % Tween 20, pH 6.0.
8. Permeabilization/wash buffer: 0.5 % Triton X-100 in PBS.
9. Blocking buffer: 5 % normal donkey serum in PBS.
10. Antibodies: antibody against UCP1 (rabbit, polyclonal, affinity purified IgG), antibody against GFP (goat polyclonal, affinity purified IgG), FITC-conjugated anti-BrdU antibody, donkey anti-rabbit Alexa Fluor 647, donkey anti-goat Alexa Fluor 594.
11. Wheat Germ Agglutinin (WGA)-Alexa Fluor 674.
12. 4′,6-diamidino-2-phenylindole (DAPI).
13. Species matched IgG control: normal rabbit IgG, normal goat IgG.

14. Aqueous mounting medium for fluorescent dye-stained tissues, 90 % Glycerol in PBS.
15. Microscope: confocal Zeizz LSM 710, Microscope Nikon ECLIPSE TS100.
16. Image analysis program: ZEN 2012.

2.3 Dissociation of Adipose Tissue and FACS Analysis of Adipocyte Isolation

1. Digestion buffer: collagenase type II 2 mg/mL, 1 % Bovine Serum Albumin (BSA) in Krebs-Ringer Bicarbonate Buffer.
2. Blocking buffer: 5 % BSA in PBS.
3. Staining buffer: 5 % BSA, 1 mM ethylenediaminetetraacetic acid (EDTA) in PBS.
4. Wash buffer: 1 % BSA in PBS.
5. Cell strainer: mesh size 70 and 100 μm.
6. Adipocyte staining reagents.
   (a) Phycoerythrin (PE) conjugated anti-mouse CD36.
   (b) HCS lipidTox™ Deep red neutral lipid stain.
   (c) Live/Dead® Viability/Cytotoxicity Kit for mammalian cells.
7. RNA isolation reagents: Trizol-LS reagent, chloroform or 1-bromo-3-chloropropane, RNA isolation kit.
8. FACS sorter (BD FACS Aria™ III).
9. FACS analysis program: FlowJo.

3 Methods

3.1 Lineage Tracing of Adipocyte Progenitors and Adipocytes During β3-Adrenergic Stimulation

1. To induce Cre recombination, double transgenic mice (Pdrgra-CreERT2/Rosa26-tdTomato, Adiponectin-CreERT2/Rosa26-mTmG) are treated with tamoxifen at the dose of 100 mg/kg by gavage for 5 days (*see* **Note 1**).
2. 10 days after the last treatment of tamoxifen, mice are infused with the β3 adrenergic receptor agonist CL316,243 (CL) at the dose of 0.75 nmol/h using mini-osmotic pumps that are subcutaneously inserted into the dorsal back region (*see* **Note 2**).
3. To trace de novo brown adipogenesis from proliferation, mice are infused with BrdU by mini-osmotic pumps (20 μg/h) (*see* **Note 3**). For cotreatment with CL using mini osmotic pumps, mix 80 mg/mL BrdU in 50 % DMSO and 3 mM CL solution or vehicle (saline) at a 1:1 ratio.

3.2 Immunohistochemical Analysis of Brown Adipocytes

Interscapular brown adipose tissue, inguinal subcutaneous white adipose tissue, and gonadal white adipose tissue, the major depots for, respectively, brown adipose tissue, subcutaneous white adipose tissue, and visceral adipose tissue of mice, are dissected for immunohistochemical analysis. Depending on the antigens of interest and target information, different types of histological preparation are required. In general, paraffin sections enable better preservation of intact morphological information, whereas frozen sections with a minimal fixation procedure have higher sensitivity to antigen detection compared with paraffin sections. UCP1 detection in histologic sections of inguinal adipose tissue from the adipocyte-tracing mouse model (adipoq-CreERT2/mTmG mice treated with CL for 7 days) will be described in this section.

3.2.1 Paraffin Sections

1. Dissected tissues are fixed in 10 % neutral buffered formalin overnight at 4 °C (*see* **Note 4**).
2. Fixed tissues are processed by standard dehydration methods or with an automatic tissue processor and embedded in paraffin.
3. Paraffin sections (5-μm thick) are prepared by using microtome (*see* **Note 5**).
4. Paraffin sections are deparaffinized and rehydrated by incubation in serial solutions of 100 % xylene, a xylene:ethanol (1:1) mixture, 100 % ethanol, 95 % ethanol, 70 % ethanol, and distilled water for 5 min each.
5. Paraffin sections are incubated with antigen-retrieval buffer in a 98 °C water bath for 10 min. The slides are washed with distilled water and then with wash buffer.
6. For BrdU detection, paraffin sections are incubated with 2 N HCl at 37 °C for 30 min, and then washed with wash buffer three times.
7. Paraffin tissue sections are blocked with blocking buffer for 15 min.
8. Paraffin sections are stained overnight with FITC-conjugated anti-BrdU antibody at 4 °C.
9. Sections are washed with wash buffer and then incubated with antibody against UCP1 and GFP at 4 °C overnight (*see* **Note 6**).
10. After washing with wash buffer twice, tissue sections are incubated with secondary antibodies conjugated with fluorophores, donkey anti-rabbit-AF647, and donkey anti-goat-AF594, for 30 min at room temperature, protected from light. DAPI in PBS (0.1 μg/L) is used for nucleus counterstaining.
11. Sections are mounted with mounting media and covered with a glass slide. After drying at room temperature, protected from light, sections are imaged with a confocal microscope. Immunostaining of the mitochondrial protein UCP1 has a punctate appearance (*see* **Note 7**; Fig. 1).

3.2.2 Cryosections

1. Tissues are fixed in 4 % paraformaldehyde for 2 h at room temperature followed by cryoprotection with a 20 % sucrose solution overnight at 4 °C.
2. Tissues are placed in Tissue Tek Cryo-OCT, mounted in a cryomold, frozen using dry ice, and stored at −80 °C or processed as 10-μm-thick cryosections.
3. The staining steps for the primary and secondary antibodies are identical to the methods described for preparation of paraffin tissue sections in Subheading 3.2.1 (*see* **Note 8**).

3.2.3 Whole Mount Staining

1. Small pieces of tissue (<2 mm^3) are fixed in 4 % paraformaldehyde for 2 h at room temperature.
2. Tissues are rinsed with PBS and then permeabilized with 1 % Triton X-100 in PBS for 1 h at room temperature.
3. Tissues are blocked with blocking buffer and incubated with primary antibody overnight at 4 °C (*see* **Note 9**).
4. The next day, the tissues are washed with PBS and incubated with the secondary antibody for 4 h at room temperature. After three washes with wash buffer, the tissue can be transferred into 50 % glycerol in PBS for 1 h and then transferred into 90 % glycerol in PBS and incubated at 37 °C for 15 min. Tissues are mounted on a cover glass slide.

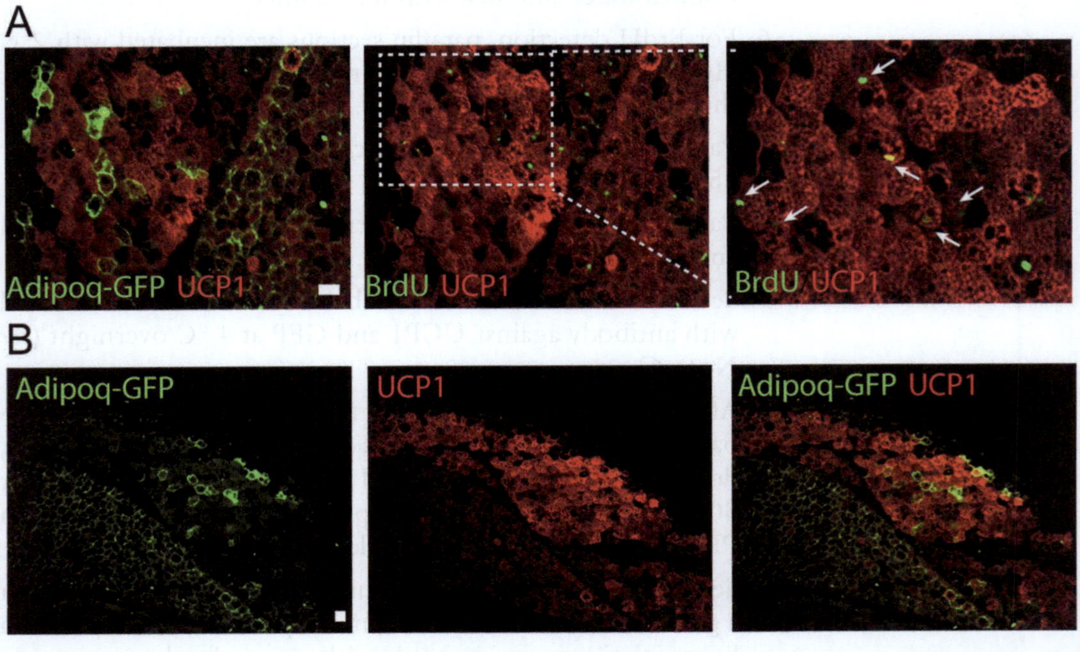

Fig. 1 UCP1 immunostaining in paraffin sections of white adipose tissue of adiponectin CreERT2/Rosa26-mTmG mice exposed to CL for 7 days. Membrane-targeted GFP expression distinguishes the cellular boundary of existing adiponectin-expressing adipocytes. High magnification (**a**) of BrdU and UCP1 staining indicates new brown adipogenesis from progenitor proliferation. (**b**) Low magnification view of UCP1 and GFP staining indicates massive de novo brown adipogenesis Bars = 20 μm

3.3 Dissociation of Adipose Tissue and FACS Analysis of Adipocytes

Isolation and analysis of single adipocytes is challenging due to their fragile nature in aqueous solution and relatively large size. A series of enzymatic/physical dissociation and isolation procedures can lead to the cell death of adipocytes. Thus, a gating strategy for single live cell sorting in combination with adipocyte markers such as lipid and CD36 expression staining is used. FACS isolation of brown adipocytes and subsequent gene expression profiling are applicable to immunophenotyping of reporter-expressing adipocytes and can be expanded by single-cell analysis technology to investigate adipocyte heterogeneity. Lee et al. [5, 10] describe a method to analyze the contribution of PDGFRα+ progenitors to adipogenesis and subsequent gene expression analysis of PDGFRα+ progenitor-derived adipocytes, using a progenitor tracing mouse model (Pdgfra-CreERT2/Rosa26-tdTomato).

3.3.1 Preparation of Dissociated Adipocytes from Adipose Tissue for Flow Cytomeric Analysis

1. Dissected adipose tissues are washed with PBS and minced using sharp scissors (1–2 mm^3).
2. Minced tissues are placed in dissociation buffer and incubated at 37 °C in a water bath with shaking for 30 min. In general, 100 mg of tissue requires 1 mL of dissociation buffer.
3. Dissociated tissues are passed through a 100-μm and a 70-μm cell strainer sequentially (*see* **Note 10**).
4. The dissociation buffer is diluted and inactivated by adding the same volume of 5 % BSA in PBS, and centrifuged at 250 × *g* for 2 min at room temperature. The top floating adipocyte fraction is collected (*see* **Note 11**). The centrifugation step is repeated to remove the contaminated fraction containing stromal vascular cells (*see* **Note 12**).
5. Floating adipocytes are stained by DAPI (0.1 μg/mL) for 1 min to exclude dead adipocytes or stained by Calcein AM (0.5 μL/mL) for 10 min at 37 °C to include live cells, and then stained for lipid as an adipocyte marker using DeepRed LipidTox staining (1:300) in staining buffer (*see* **Note 13**).
6. After staining, the floating adipocyte fraction is washed with wash buffer, centrifugated at 250 × *g* for 2 min, collected, and resuspended in staining buffer.
7. Flow cytometric analysis of lipid+ Pdgfra-tdTomato+ adipocytes is performed to determine the contribution of progenitors to de novo adipogenesis (Fig. 2).

3.3.2 FACS-Isolation of Live Adipocytes for Subsequent RNA Analysis

1. To prepare the collection tubes for the isolated adipocytes, 5 mL FACS tubes are filled with 3 mL of TRIzol LS reagent and 1 mL samples are collected to yield approximately 2 × 10^5 cells/mL per tube (*see* **Note 14**).
2. Up to 1 mL of the cells are collected, then vortexed and frozen at −80 °C.

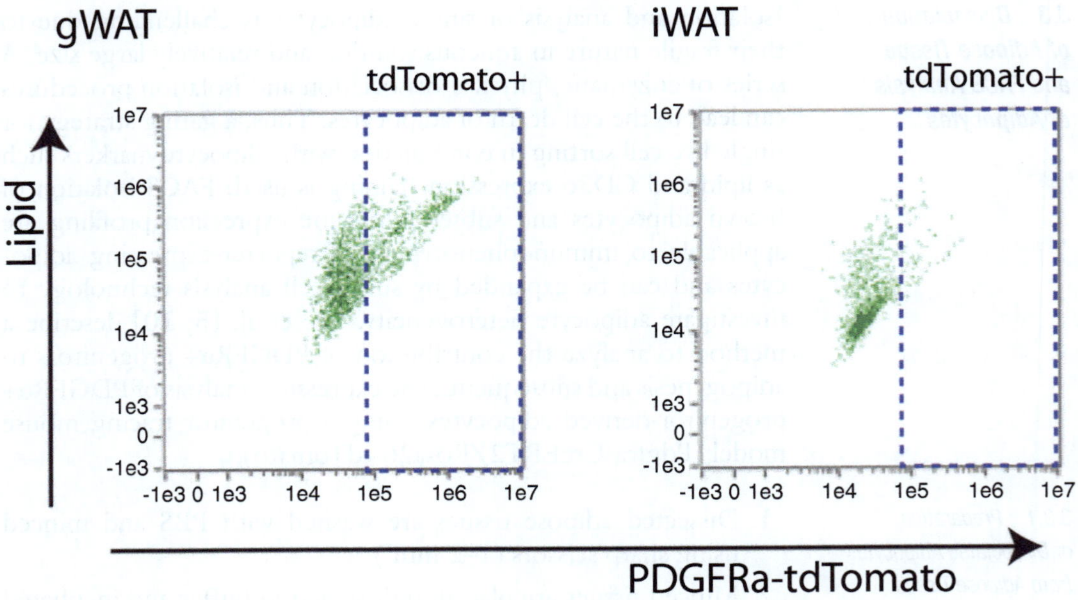

Fig. 2 Flow cytometric analysis of dissociated tdTomato+ adipocytes from PDGFRaCreERT2/Rosa26-tdTomato mice exposed to CL for 7 days. Flow cytometric analysis of lipid+ adipocytes from gonadal white adipose tissue (gWAT) and inguinal white adipose tissue (iWAT) indicates different levels of tdTomato expression

3. Samples are subjected to RNA purification using RNA isolation kit and quantitative PCR analysis for brown adipocyte marker expression.

4 Notes

1. The tamoxifen dose should be determined for each strain because the sensitivity to tamoxifen and efficiency of recombination can vary depending on the lineage marking system. For clonal analysis, a low dose can be used. Recombination efficiency and specificity after tamoxifen treatment must be determined for each cell type.
2. One day of CL infusion can induce UCP1+ brown adipocytes in inguinal subcutaneous adipose tissue detectable by histological and western blot analysis. While UCP1 levels in gonadal adipose tissue vary depending on the mouse strain, 7 days of CL infusion can induce UCP1+ brown adipocytes in gonadal adipose tissue of 129S1 mice. In addition to pharmacological activation, cold exposure is a physiological thermogenic stimuli and mainly relies on sympathetic activation; thus, the level of innervation of adipose tissue depots is one of the important

factors for data interpretation. Treatment with beta adrenergic receptor agonist with surgical and chemical denervation of adipose tissue has been useful to dissect the neurologic effects of sympathetic innervation from adrenergic receptor activation in adipocytes [10].

3. For histological detection of BrdU in adipose tissue, the intestine is dissected from the same individual mice as the positive controls for thymidine analog detections in histological sections.
4. A fixation time longer than 48 h is not recommended.
5. Paraffin can be melt at 65 °C before staining.
6. For blocking, serum from the same species of origin as the secondary antibodies are recommended. Alternatively, 5 % BSA in PBS solution can be used. For negative controls, use normal serum at the same concentration of primary antibody.
7. Mitochondrial localization of UCP1 can be confirmed by double staining with anti-prohibitin antibody. To visualize the morphology of multilocular adipocytes, double immunostaining for UCP1 and perilipin 1 (PLIN1) or H/E staining of adjacent sections is useful and informative as an additional indicator of the brown adipocyte phenotype.
8. Using ice-cold reagents helps maintain lipid droplet integrity cryosections.
9. It is important to use 0.5 % Triton X-100 during antibody incubation in whole-mount staining.
10. In general, brown adipocytes (~ 40 μm) are relatively smaller than white adipocytes. The large adipocyte fraction from white adipose tissue cannot pass through a 40-μm cell strainer.
11. For transferring the adipocyte fraction, a glass pipet is useful, minimizing the loss of adipocytes that aggregate on plastic pipet tips.
12. Centrifugation at room temperature is recommended because low temperature (e.g., 4 °C) causes aggregation of dissociated adipocytes, resulting in clumps.
13. For an alternative adipocyte marker, anti-Cd36 antibody can be used.
14. The number of cells collected will depend on the sorting conditions; however, do not exceed a 1 mL sample volume per 3 mL of TRIzol LS reagent. The purity of FACS-isolated adipocytes can be confirmed by testing for enrichment of adipogenic gene expression in FACS-isolated cells compared with expression levels in floating fractions that are isolated only by centrifugation.

Acknowledgments

We thank Drs. Granneman, Sanders, Kim, and Kwon for constructive comments on the manuscript. This work is supported by National Research Foundation of Korea grant NRF-2014R1A6A3A04056472 (Y.H.L.).

References

1. Kajimura S, Spiegelman Bruce M, Seale P (2015) Brown and Beige fat: physiological roles beyond heat generation. Cell Metab 22(4):546–559
2. Saito M, Okamatsu-Ogura Y, Matsushita M, Watanabe K, Yoneshiro T, Nio-Kobayashi J, Iwanaga T, Miyagawa M, Kameya T, Nakada K, Kawai Y, Tsujisaki M (2009) High incidence of metabolically active brown adipose tissue in healthy adult humans: effects of cold exposure and adiposity. (ORIGINAL ARTICLE)(Report). Diabetes 58(7): 1526–1531
3. Cypess AM, Lehman S, Williams G, Tal I, Rodman D, Goldfine AB, Kuo FC, Palmer EL, Tseng YH, Doria A, Kolodny GM, Kahn CR (2009) Identification and importance of brown adipose tissue in adult humans. N Engl J Med 360(15):1509–1517
4. Wu J, Boström P, Sparks Lauren M, Ye L, Choi Jang H, Giang A-H, Khandekar M, Virtanen Kirsi A, Nuutila P, Schaart G, Huang K, Tu H, van Marken Lichtenbelt Wouter D, Hoeks J, Enerbäck S, Schrauwen P, Spiegelman Bruce M (2012) Beige adipocytes are a distinct type of thermogenic fat cell in mouse and human. Cell 150(2):366–376
5. Lee Y-H, Petkova Anelia P, Mottillo Emilio P, Granneman James G (2012) In vivo identification of bipotential adipocyte progenitors recruited by β3-adrenoceptor activation and high-fat feeding. Cell Metab 15(4):480–491
6. Long Jonathan Z, Svensson Katrin J, Tsai L, Zeng X, Roh Hyun C, Kong X, Rao Rajesh R, Lou J, Lokurkar I, Baur W, Castellot Jr John J, Rosen Evan D, Spiegelman Bruce M (2014) A smooth muscle-like origin for beige adipocytes. Cell Metab 19(5):810–820
7. Wang QA, Tao C, Gupta RK, Scherer PE (2013) Tracking adipogenesis during white adipose tissue development, expansion and regeneration. Nat Med 19(10):1338–1344
8. Kretzschmar K, Watt Fiona M (2012) Lineage Tracing. Cell 148(1):33–45
9. Berry R, Rodeheffer MS (2013) Characterization of the adipocyte cellular lineage in vivo. Nat Cell Biol 15(3):302–308
10. Lee YH, Petkova AP, Konkar AA, Granneman JG (2015) Cellular origins of cold-induced brown adipocytes in adult mice. FASEB J 29(1):286–299
11. Lee Y-H, Petkova Anelia P, Granneman James G (2013) Identification of an adipogenic niche for adipose tissue remodeling and restoration. Cell Metab 18(3):355–367
12. Rosenwald M, Perdikari A, Rülicke T, Wolfrum C (2013) Bi-directional interconversion of brite and white adipocytes. Nat Cell Biol 15(6):659–667
13. Lee Y-H, Thacker RI, Hall BE, Kong R, Granneman JG (2014) Exploring the activated adipogenic niche: Interactions of macrophages and adipocyte progenitors. Cell Cycle 13(2):184–190
14. Zawadzka M, Rivers LE, Fancy SPJ, Zhao C, Tripathi R, Jamen F, Young K, Goncharevich A, Pohl H, Rizzi M, Rowitch DH, Kessaris N, Suter U, Richardson WD, Franklin RJM (2010) CNS-resident glial progenitor/stem cells produce Schwann cells as well as oligodendrocytes during repair of CNS demyelination. Cell Stem Cell 6(6):578–590

Chapter 10

Genetic Mouse Models: The Powerful Tools to Study Fat Tissues

Xingxing Kong, Kevin W. Williams, and Tiemin Liu

Abstract

Obesity and Type 2 diabetes (T2D) are associated with a variety of comorbidities that contribute to mortality around the world. Although significant effort has been expended in understanding mechanisms that mitigate the consequences of this epidemic, the field has experienced limited success thus far. The potential ability of brown adipose tissue (BAT) to counteract obesity and metabolic disease in rodents (and potentially in humans) has been a topical realization. Recently, there is also another thermogenic fat cell called beige adipocytes, which are located among white adipocytes and share similar activated responses to cyclic AMP as classical BAT. In this chapter, we review contemporary molecular strategies to investigate the role of adipose tissue depots in metabolism. In particular, we will discuss the generation of adipose tissue-specific knockout and overexpression of target genes in various mouse models. We will also discuss how to use different Cre (cyclization recombination) mouse lines to investigate diverse types of adipocytes.

Key words Adiponectin-cre, Ucp1-cre, Beige adipocyte, Rosa26-loxp-stop-loxp

1 Introduction

Adipocytes can be broadly classified as white or brown fat cells. While white fat cells are specialized to store energy in the form of triglyceride, brown adipocytes utilize their high mitochondrial content and uncoupling protein 1 (UCP1) to uncouple respiration and dissipate chemical energy as heat [1]. Rodents and other small mammals have copious brown fat deposits, but larger mammals often lose brown fat after infancy [2]. Recent data indicates that adult humans contain significant deposits of UCP1-positive brown fat that can be detected by PET-CT scanning methods [3–7]. Another type of adipocyte has been identified, which are "brown-like" cells that express UCP1 after cold exposure, but which appear within white depots rather than the classic interscapular BAT. These "inducible" cells have been called "beige" or "brite" cells [8–11]. The characteristics of BAT as an energy-burning tissue have led to interest in harnessing this activity to combat obesity.

Jun Wu (ed.), *Thermogenic Fat: Methods and Protocols*, Methods in Molecular Biology, vol. 1566,
DOI 10.1007/978-1-4939-6820-6_10, © Springer Science+Business Media LLC 2017

One might accomplish this by stimulating the activity of already existing BAT, or perhaps by inducing the appearance of new beige cells.

In order to characterize the biology among the three types of adipocytes, we will discuss the specific mouse models for investigating different fat depots.

2 Materials

2.1 Strategies for Generating Adipocyte-Specific Cre Mouse Line

1. Adiponectin BAC (90G21).
2. UCP1 BAC (148M1).
3. EL250 cell, Cre recombinase cassette with a Frt-flanked neomycin resistance cassette.
4. LA Taq DNA Polymerase (TaKaRa).
5. QIAGEN Plasmid Midiprep Kit.

2.2 Strategies for Generating Adipose Tissue Transgenic Mouse Line

1. Taq polymerase.
2. Mlu I and Nsi I enzymes and buffer.
3. T4 ligase.
4. 2.2.4 KpnI enzyme.
5. 1.0 % agarose gel.
6. Ultracentrifuge.
7. 1.7-mL microcentrifuge tubes.
8. Microcapillary pipet.
9. DNA extraction kit (Qiagen).
10. Elutip-D column.
11. Additional reagents and equipment for checking the DNA concentration using a spectrophotometer.
12. Digestion buffer: 10 mM Tris–HCl (pH 7.6–8.0); 25 mM EDTA; 100 Mm NaCl; 0.5 % SDS.
13. Proteinase K.

3 Methods

3.1 Strategies for Generating Adipocyte-Specific Cre Mouse Line (Adiponectin-Cre as an Example)

Adiponectin promoter driving Cre expression (Adiponectin-cre) in white adipose tissue and brown adipose tissue (*see* **Notes 1** and **2**).

1. Electroporate adiponectin BAC (90G21) into EL250 cells.
2. Using PCR, amplify a linear dsDNA Cre recombinase cassette with a Frt-flanked neomycin resistance cassette that includes 70 bp of DNA with homologous flanking regions to adiponectin gene.

3. Electroporate the above PCR amplified Cre recombinase cassette into EL250 cells containing adiponectin BAC.
4. Use Luria-Bertani (LB) agar plates containing neomycin (40 μg/mL Kanamycin) to select positive targeting cassette into specific adiponectin region by homologous recombination.
5. Pick up the single positive colony and grow in 30 mL LB medium containing 40 μg/mL Kanamycin overnight.
6. Obtain positive BAC DNA using QIAGEN Plasmid Midiprep Kit, then sequence the whole Cre recombinase cassette.
7. Remove the Frt-flanked neomycin resistance cassette from the correct sequence of adiponectin BAC DNA containing Cre recombinase cassette [12].
8. Inject the adiponectin BAC DNA containing Cre recombinase cassette into fertilized mouse eggs.
9. Each founder line is crossed with the R26R tomato reporter line to map the activity pattern of Cre recombinase, then select the Adiponectin-cre founder line that expresses Cre recombinase-activated tomato in the greatest number of adiponectin cells. One example followed in the next part [12].

3.2 Strategies for Generating Adipose Tissue Transgenic Mouse Line (Rosa26-Transgenic Mice)

Strategies to overexpress genes in adipocytes using the aP2 (fatty acid binding protein 4, FABP4, commonly known as adipocyte Protein 2) promoter or Adiponectin promoter followed by the target genes. As previously mentioned by Kang et al., this method has its limitations, for example, variation in transgene copy number and position effects [13]. Another strategy for overexpression of genes in specific tissues is through the use of cre and loxp-stop-loxp system. Several studies have demonstrated the value of using a cre/loxp strategy; however, we are illustrating the Rosa26 targeting approach. Specifically, the transgene is cloned into the ROSA26 locus downstream of a loxP-flanked stuffer DNA sequence (STOP cassette), which abrogates transgene expression [14]. The authors modified the plasmid as shown in Fig. 1. Upon temporal and cell-type-specific induction of Cre, transcription of the transgene from the ROSA26 promoter (or from an exogenous one inserted into the ROSA26 locus) is induced as a result of the deletion of the loxP-flanked STOP cassette. To trace expression of the transgene in vivo, it can be useful to introduce downstream of the transgene an Internal Ribosome Entry Site (IRES) followed by a reporter gene coding for a fluorescence protein (such as GFP) or an enzyme (β-galactosidase) (*see* **Note 3**). However, in our case we used 2A peptide instead of IRES to increase the cleavage efficiency between genes upstream and downstream of it.

3.2.1 Construct Design and Purification

Every step (from designing the transgenic construct to embryo transfer) is critical to successfully generate transgenic mice. First,

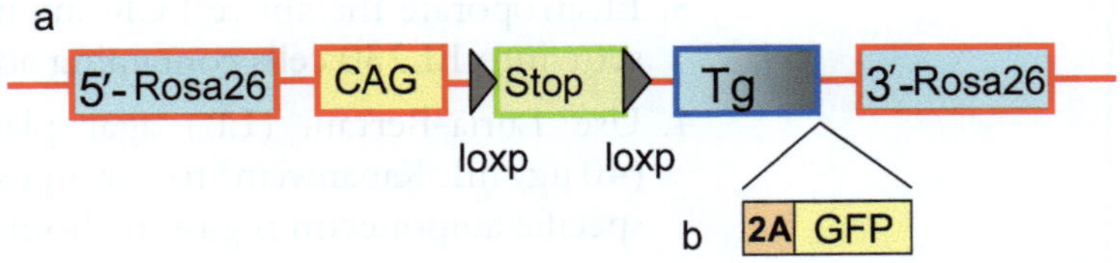

Fig. 1 Targeting strategy to insert transgene into Rosa26 locus. Conditional transgene expressed from the CAG promoter upon Cre-mediated recombination (**a**). To monitor expression of the transgene in vivo a 2A-GFP cassette can be cloned downstream of the transgene of interest (**b**). 2A is a self-cleaving peptide, having high cleavage efficiency between genes upstream and downstream of it

preparing a clean DNA sample is a vital step because it affects the health of the embryo and the DNA integration efficiency. A DNA fragment lacking any trace of vector sequence should only be microinjected. The vector sequence, as well as any chemical residues remaining in the final DNA solution, is generally toxic to mouse zygotes and will result in death of the embryo or poor efficiency in generating transgenic mice.

1. Amplify the target gene from the cDNA and use restriction enzymes and DNA ligases to cut and rejoin DNA fragments that in our case we use Mlu I and Nsi I (Fig. 2).
2. Linearize the cloned cassette. In our example, KpnI is used to linearize the plasmid.
3. DNA purification. We routinely use either of two methods to purify DNA (Elutip-D or a gel-purification method). Both methods have their own merits in that the Elutip-D method yields the purest DNA for microinjection, whereas the gel purification method is a quick and easy method to yield sufficiently DNA for microinjection. (1) Elutip-D method: use phenol: chloroform to extract and EtOH precipitate the digested DNA. Use the Elutip-D column to purify DNA. (2) Gel-purify the construct: Run the digested plasmid DNA transgenic construct on a 0.8 % agarose gel in 1× TAE buffer. Excise the transgene DNA band from the agarose gel with a clean razor blade. Use the Qiagen gel extraction kit to extract DNA.
4. Measure the DNA concentration. Most injection core facilities require at least 90 μg DNA to inject into ES cells.
5. Microinjection. DNA was injected into 129Sv embryonic stem cells (which are most commonly used) as performed by the core facility.

3.2.2 ES Cells Selection

Before submitting the ES cells for microinjection, collect the ES cell DNA to identify the transgene within the ES cells (Southern blot or long-range PCR). The 5′ primers are 5′-GCCAAGTGGGCAGTTTACCG-3′ and 5′-TAGGTAGG

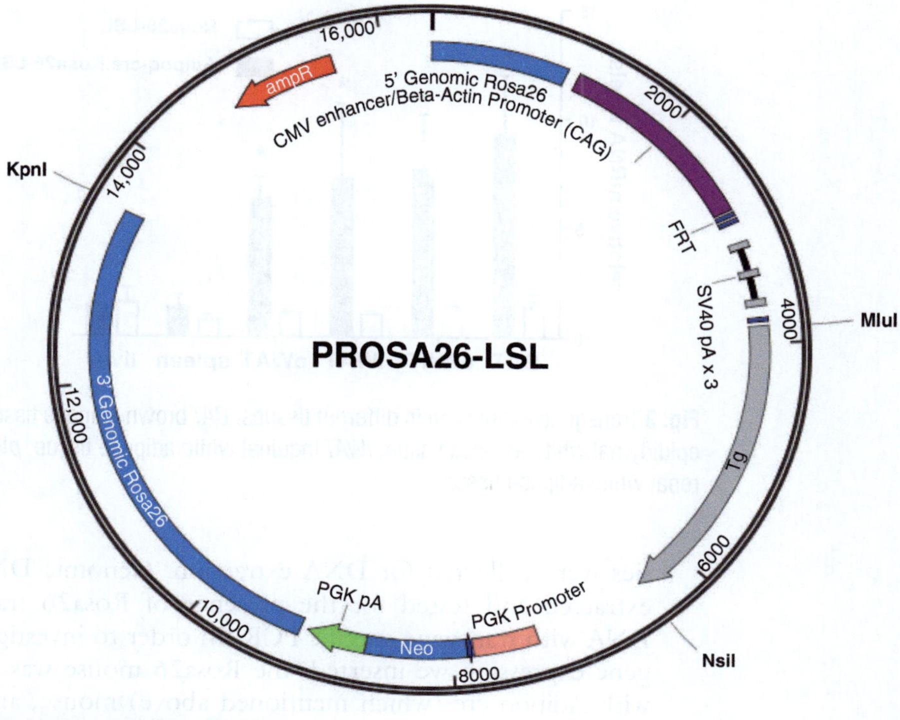

Fig. 2 The construct of pRosa26-LSL

GGATCGGGACTCT-3′. The 3′ primers are 5′-GCCAGCTCATTCCTCCCACTC-3′ and 5′-GGCATGGCAATGTTCAAGCAG-3′.

1. Culture ES cells until confluency in a 24-well plate.
2. Aspirate the medium and add 400 μL of digestion buffer and 5 μL of 20 mg/mL proteinase K.
3. Incubate in 60 °C shaker, overnight.
4. Add equal volume of Phenol/Chlo, vortex (mix) well.
5. Extract and precipitate with 100 % cold ethanol (not add salt).
6. Wash twice with 70 % ethanol.
7. Resuspend in 30–40 μL of TE depending on the size of pellet.
8. Measure the concentration of DNA.
9. Do long-range PCR by using the primers listed above or store the samples in −20 °C.
10. ES cells microinjection. Intact microinjected eggs were transferred to the oviducts of pseudo-pregnant recipients by the core.
11. Generation of chimeric mice and identification of germ-line transmission. Ear tags or punches were applied to 2-week-old mouse pups for identification purposes and 5–10 mm tail biop-

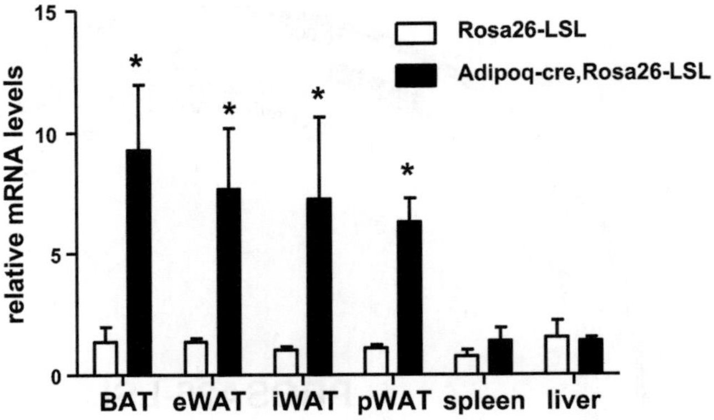

Fig. 3 Transgene expression in different tissues. *BAT* brown adipose tissue, *eWAT* epididymal white adipose tissue, *iWAT* inguinal white adipose tissue, *pWAT* perirenal white adipose tissue

sies were collected for DNA extraction. Genomic DNA was extracted and tested for the presence of Rosa26 transgene DNA with transgene specific PCR. In order to investigate the gene expression we inserted, the Rosa26 mouse was crossed with Adipoq-cre (which mentioned above) mouse, and then detect the gene expression in different fat tissues (Fig. 3) (*see* **Note 4**).

4 Notes

1. To achieve adipose-specific gene targeting, several Cre lines have been generated with the use of the adipose-specific gene regulatory element [15–23]. Tissue specificity and efficacy of recombination using the Cre/LoxP (cyclization recombination-locus of X over P1) system are affected by many parameters, including the genomic location and distance between the loxP sites of the target locus as well as insertion site and copy number of Cre transgene [24], reviewed in detail by Kang S. [13].
2. So far, the field currently lacks a Cre line that targets WAT-specific or beige-specific recombination. The Adiponectin-cre targets all fat depots while the Ucp1-cre is restricted to BAT. However, given that beige adipocytes can be driven to upregulate the expression of Ucp1 by cold and other stimuli, Ucp1-cre recombination would occur in depots outside classical BAT. Importantly, Wu and Spiegelman have identified markers selectively for beige cells, such as CD137, CD40, and TMEM26 [2]. Thus, it will be at least theoretically possible to generate a beige-specific Cre mouse line.

3. The traditional transgenic mice are usually generated by direct microinjection of DNA fragments into fertilized one-cell mouse embryos, followed by transfer of these embryos into recipient mothers that can carry the pregnancy to term. Our strategy is gene targeting approach requires embryonic stem cells (ES cells), which are currently only available for a very few strains of mice (i.e., 129Sv and C57/bL6J). However, the major advantage of using Rosa26 for transgenic studies is that it can overcome positional effects to produce integration-site independent, copy-number dependent, and accurate transgene expression in vivo. Importantly, the *Rosa26* locus itself can be targeted with relative ease and shows broad expression across most cell types. Furthermore, contribute to the loxp-stop-loxp site, the target gene can be expressed in a time- and cell-type specific fashion.
4. Finally, the Cre/loxP system can control transgene expression in a time- and cell-type specific fashion. However, once induced, the transgene cannot be silenced any longer. To overcome this limitation, the tetracycline (Tet)-controlled system can be applied to generate inducible ROSA26 transgenes [25].

Reference

1. Rosen ED, Spiegelman BM (2014) What we talk about when we talk about fat. Cell 156(1–2):20–44. doi:10.1016/j.cell.2013.12.012 S0092-8674(13)01546-8 [pii]. Epub 2014/01/21. PubMed PMID: 24439368; PubMed Central PMCID: PMC3934003
2. Wu J, Bostrom P, Sparks LM, Ye L, Choi JH, Giang AH et al (2012) Beige adipocytes are a distinct type of thermogenic fat cell in mouse and human. Cell 150(2):366–376. doi:10.1016/j.cell.2012.05.016 S0092-8674(12)00595-8 [pii]. Epub 2012/07/17. PubMed PMID: 22796012; PubMed Central PMCID: PMC3402601
3. Cypess AM, Lehman S, Williams G, Tal I, Rodman D, Goldfine AB et al (2009) Identification and importance of brown adipose tissue in adult humans. N Engl J Med 360(15):1509–1517. doi:10.1056/NEJMoa0810780 360/15/1509 [pii]. Epub 2009/04/10. PubMed PMID: 19357406; PubMed Central PMCID: PMC2859951
4. Mirbolooki MR, Constantinescu CC, Pan ML, Mukherjee J (2011) Quantitative assessment of brown adipose tissue metabolic activity and volume using 18F-FDG PET/CT and β3-adrenergic receptor activation. EJNMMI Res 1(1):30. doi:10.1186/2191-219X-1-30 2191-219X-1-30 [pii]. Epub 2012/01/05. PubMed PMID: 22214183; PubMed Central PMCID: PMC3250993
5. Virtanen KA, Lidell ME, Orava J, Heglind M, Westergren R, Niemi T et al (2009) Functional brown adipose tissue in healthy adults. N Engl J Med 360(15):1518–1525. doi:10.1056/NEJMoa0808949 360/15/1518 [pii]. Epub 2009/04/10. PubMed PMID: 19357407
6. von Heydebreck A, Huber W, Poustka A, Vingron M (2001) Identifying splits with clear separation: a new class discovery method for gene expression data. Bioinformatics 17(Suppl 1):S107–S114 Epub 2001/07/27. PubMed PMID: 11472999
7. Orava J, Nuutila P, Lidell ME, Oikonen V, Noponen T, Viljanen T et al (2011) Different metabolic responses of human brown adipose tissue to activation by cold and insulin. Cell Metab 14(2):272–279. doi:10.1016/j.cmet.2011.06.012 S1550-4131(11)00261-0 [pii]. Epub 2011/08/02. PubMed PMID: 21803297
8. Seale P, Conroe HM, Estall J, Kajimura S, Frontini A, Ishibashi J et al (2011) Prdm16 determines the thermogenic program of

subcutaneous white adipose tissue in mice. J Clin Invest 121(1):96–105. doi:10.1172/JCI44271 44271 [pii]. Epub 2010/12/03. PubMed PMID: 21123942; PubMed Central PMCID: PMC3007155

9. Petrovic N, Walden TB, Shabalina IG, Timmons JA, Cannon B, Nedergaard J (2010) Chronic peroxisome proliferator-activated receptor gamma (PPARgamma) activation of epididymally derived white adipocyte cultures reveals a population of thermogenically competent, UCP1-containing adipocytes molecularly distinct from classic brown adipocytes. J Biol Chem 285(10):7153–7164. doi:10.1074/jbc.M109.053942 M109.053942 [pii]. Epub 2009/12/24. PubMed PMID: 20028987; PubMed Central PMCID: PMC2844165

10. Ishibashi J, Seale P (2010) Medicine. Beige can be slimming. Science 328(5982):1113–1114. doi:10.1126/science.1190816 science.1190816 [pii]. Epub 2010/05/08. PubMed PMID: 20448151; PubMed Central PMCID: PMC2907667

11. Seale P, Bjork B, Yang W, Kajimura S, Chin S, Kuang S et al (2008) PRDM16 controls a brown fat/skeletal muscle switch. Nature 454(7207):961–967. doi:10.1038/nature07182 nature07182 [pii]. Epub 2008/08/23. PubMed PMID: 18719582; PubMed Central PMCID: PMC2583329

12. Dhillon H, Zigman JM, Ye C, Lee CE, McGovern RA, Tang V et al (2006) Leptin directly activates SF1 neurons in the VMH, and this action by leptin is required for normal body-weight homeostasis. Neuron 49(2):191–203 PubMed PMID: 16423694

13. Kang S, Kong X, Rosen ED (2014) Adipocyte-specific transgenic and knockout models. Methods Enzymol 537:1–16. doi:10.1016/B978-0-12-411619-1.00001-X B978-0-12-411619-1.00001-X [pii]. Epub 2014/02/01. PubMed PMID: 24480338

14. Soriano P (1999) Generalized lacZ expression with the ROSA26 Cre reporter strain. Nat Genet 21(1):70–71. doi:10.1038/5007 Epub 1999/01/23. PubMed PMID: 9916792

15. Graves RA, Tontonoz P, Platt KA, Ross SR, Spiegelman BM (1992) Identification of a fat cell enhancer: analysis of requirements for adipose tissue-specific gene expression. J Cell Biochem 49(3):219–224. doi:10.1002/jcb.240490303 Epub 1992/07/01. PubMed PMID: 1644859

16. Ross SR, Graves RA, Greenstein A, Platt KA, Shyu HL, Mellovitz B et al (1990) A fat-specific enhancer is the primary determinant of gene expression for adipocyte P2 in vivo. Proc Natl Acad Sci U S A 87(24):9590–9594 Epub 1990/12/01. PubMed PMID: 2263614; PubMed Central PMCID: PMC55218

17. Barlow C, Schroeder M, Lekstrom-Himes J, Kylefjord H, Deng CX, Wynshaw-Boris A et al (1997) Targeted expression of Cre recombinase to adipose tissue of transgenic mice directs adipose-specific excision of loxP-flanked gene segments. Nucleic Acids Res 25(12):2543–2545 doi: gka413 [pii]. Epub 1997/06/15. PubMed PMID: 9171115; PubMed Central PMCID: PMC146759

18. Abel ED, Peroni O, Kim JK, Kim YB, Boss O, Hadro E et al (2001) Adipose-selective targeting of the GLUT4 gene impairs insulin action in muscle and liver. Nature 409(6821):729–733. doi:10.1038/35055575 Epub 2001/02/24. PubMed PMID: 11217863

19. He W, Barak Y, Hevener A, Olson P, Liao D, Le J et al (2003) Adipose-specific peroxisome proliferator-activated receptor gamma knockout causes insulin resistance in fat and liver but not in muscle. Proc Natl Acad Sci U S A 100(26):15712–15717. doi:10.1073/pnas.2536828100 2536828100 [pii]. Epub 2003/12/09. PubMed PMID: 14660788; PubMed Central PMCID: PMC307633

20. Eguchi J, Wang X, Yu S, Kershaw EE, Chiu PC, Dushay J et al (2011) Transcriptional control of adipose lipid handling by IRF4. Cell Metab 13(3):249–259. doi:10.1016/j.cmet.2011.02.005 Epub 2011/03/02. PubMed PMID: 21356515; PubMed Central PMCID: PMC3063358

21. Imai T, Jiang M, Chambon P, Metzger D (2001) Impaired adipogenesis and lipolysis in the mouse upon selective ablation of the retinoid X receptor alpha mediated by a tamoxifen-inducible chimeric Cre recombinase (Cre-ERT2) in adipocytes. Proc Natl Acad Sci U S A 98(1):224–228. doi:10.1073/pnas.011528898 011528898 [pii]. Epub 2001/01/03. PubMed PMID: 11134524; PubMed Central PMCID: PMC14572

22. Wang ZV, Deng Y, Wang QA, Sun K, Scherer PE (2010) Identification and characterization of a promoter cassette conferring adipocyte-specific gene expression. Endocrinology 151(6):2933–2939. doi:10.1210/en.2010-0136 en.2010-0136 [pii]. Epub 2010/04/07.

PubMed PMID: 20363877; PubMed Central PMCID: PMC2875825

23. Kong X, Banks A, Liu T, Kazak L, Rao RR, Cohen P et al (2014) IRF4 is a key thermogenic transcriptional partner of PGC-1alpha. Cell 158(1):69–83. doi:10.1016/j.cell.2014.04.049 S0092-8674(14)00723-5 [pii]. Epub 2014/07/06. PubMed PMID: 24995979; PubMed Central PMCID: PMC4116691

24. Kos CH (2004) Cre/loxP system for generating tissue-specific knockout mouse models. Nutr Rev 62(6 Pt 1):243–246 Epub 2004/08/05. PubMed PMID: 15291397

25. Beard C, Hochedlinger K, Plath K, Wutz A, Jaenisch R (2006) Efficient method to generate single-copy transgenic mice by site-specific integration in embryonic stem cells. Genesis 44(1):23–28. doi:10.1002/gene.20180 Epub 2006/01/10. PubMed PMID: 16400644

Chapter 11

Genetic Manipulation with Viral Vectors to Assess Metabolism and Adipose Tissue Function

Nicolás Gómez-Banoy and James C. Lo

Abstract

Viral vectors have become widely used tools for genetic manipulation of adipose tissues to understand the biology and function of adipocytes in metabolism. There are a number of different viral vectors commonly used: retrovirus, lentivirus, adenovirus, and adeno-associated virus (AAV). Here, we review examples from the literature and describe methods to transduce adipocytes and adipose tissues using retrovirus, lentivirus, adenovirus, and AAV to ascertain gene function in adipose biology.

Key words Adipose tissue, Brown fat, White fat, Beige fat, Adipocyte, Viral vectors, Adenovirus, Transduction, Retrovirus, Lentivirus, Adeno-associated virus, AAV, Adipogenesis, Adipokine, Metabolism, Diabetes, Obesity

1 Introduction to Adipose

Interest in adipose tissue function has come a long way in the last few decades [1]. Adipose tissue was once thought to be solely an energy storage organ. It in fact is probably the best energy storage organ that mammals possess. In times of energy surplus, organisms that possess adipose tissue or a similar organ such as the fat body in the model organism *Drosophila melanogaster* can bank their surplus energy in the form of fat [2, 3]. The adipocyte plays a central role in lipid metabolism through lipogenesis and lipolysis. The dysregulation of adipocyte lipid metabolism is a major driver of the pathogenesis of metabolic diseases such as obesity and type 2 diabetes [4, 5].

After extensive work on the basic biochemistry of lipid metabolism within adipocytes had been elucidated, interest in the field of adipocyte biology initially waned. Later, it became apparent that adipose tissue serves other functions besides energy storage, spurring further interest in the area of adipose. Adipocytes were much more active than previously thought and have the ability to affect systemic metabolic processes beyond energy storage. It was noted

Jun Wu (ed.), *Thermogenic Fat: Methods and Protocols*, Methods in Molecular Biology, vol. 1566,
DOI 10.1007/978-1-4939-6820-6_11, © Springer Science+Business Media LLC 2017

that adipocytes secrete soluble factors that were dubbed "adipocytokines" or "adipokines." These adipokines can work at many different levels. Adipokines can function at an autocrine level acting on the same cell that secreted it. Adipokines can also stimulate other cells in the local environment in a paracrine fashion. Lastly, adipokines can enter the circulation to have systemic effects on distant tissues such as leptin signaling to the brain or TNF signaling to the many different cell types in the body that express TNF receptors. Other hormones such as insulin communicate back to the adipose tissue completing the organ cross talk.

1.1 Introduction to Viral Vectors

Over the last five decades, the fields of virology and molecular biology have exploded and huge strides have been made in discovering new viruses and understanding their life cycles. As we learned more about the properties of viruses and the necessary components and proteins responsible for replication, host cell pathogenicity, infection, and immune evasion, viral vectors were developed that could efficiently modulate gene expression of a target cell. There have now been several generations of many of the popular viral vectors, taking advantage of the rapidly acquired knowledge to provide higher efficiency, ease of use, and an enhanced safety profile.

In this section, we will review some of the more commonly used viral vectors in modern day biology and how they have furthered our understanding of adipose tissue function and the general field of molecular metabolism. We will go over the advantages and disadvantages of these vectors and reference examples to illustrate their utility in adipocyte research. Here in Table 1 we summarize the basic features of the viral vectors discussed in this chapter.

1.2 Retroviruses

The viral family of retroviruses (*Retroviridae*) includes three major subfamilies: lentiviruses, alpha-retroviruses, and gamma-retroviruses. The main feature of the retrovirus family is the utilization of the viral enzymes reverse transcriptase (RT) and integrase (IN) for the insertion of genomic information into the host genome. There is high similarity in the structure, life cycle, and main proteins of the three subfamilies. In general, all retroviral-based vectors are appealing models for genome editing due to their capacity of integrating stably into the host genome and achieving long-term transduction [6]. Lentiviruses possess a more complex genome due to the expression of accessory proteins and more complex transcriptional units. The main feature that distinguishes them from other retroviruses is their capacity to transduce nondividing cells, which makes them appealing candidates for gene therapy in a variety of tissues and cell lines. The human immunodeficiency viruses 1 and 2 (HIV-1 and 2) and the simian immunodeficiency virus (SIV) have been the model for developing lentivirus-based gene therapy [7]. In this chapter, we will begin by outlining the

Table 1
General characteristics of viral vectors for adipose tissue transduction

	Lentivirus	Retrovirus	Adenovirus	Adeno-associated virus
Viral genome	Single-stranded RNA	Single-stranded RNA	Double-stranded DNA	Single-stranded DNA
Transduction of nondividing cells	Positive	Negative	Positive	Positive
Tropism	$CD4^+$ T cells. May expand its tropism if pseudotyped with VSG-G	Immune cells. May expand its tropism if pseudotyped with VSG-G	Hepatocytes	Depending on serotype (12 human serotypes AAV1-12)
Integration into host genome	Yes	Yes	No	No
Duration of expression	Weeks/months	Weeks/months	Days—2 weeks	Weeks/months
Safety concerns	Insertional mutagenesis	Insertional mutagenesis	Exaggerated immune/inflammatory response	Immune response
BLS category	2	2	1/2	1

use of lentiviral vectors in adipose tissue genetic engineering. Afterward, we will discuss other retroviral-based vectors and their utility in adipocyte genome editing.

1.3 Lentiviral Vectors and Adipose Tissue

The transition of HIV-1, a tremendously pathogenic virus for human, into a transgene-carrying virus has undergone multiple steps, in which the main concern is biological safety. The generation of replication competent recombinants (RCR) and potentially pathological particles for the host is the major issue when using replication-defective gene delivery vectors. Thus, in order to reduce the probability of generating RCRs, lentiviral-based vectors have undergone a series of transformations throughout the years, leading to first, second, and third generation models [8].

Lentiviral-based vectors have been used for genetic engineering of adipocytes in vitro in human and animal models. One of the first publications on the subject compared adenoviral, oncoretroviral, and lentiviral models for genetic transduction in mesenchymal progenitor (PLA) cells derived from adipose tissue. In contrast to oncoretroviruses, lentiviruses transduced mouse embryonic stem cells (ES) and most importantly PLA cells derived from human adipose tissue with high efficiency. The transduced cells retained persistent EGFP expression even after differentiation into adipocytes and osteogenic cells. This demonstrated that, in comparison with adenoviral and oncoretroviral models, lentiviral vectors could achieve stable gene expression after long-term culture and differentiation [9]. Other early studies in vitro investigated the transduction efficiency of lentiviral-based models in mature 3T3-L1 adipocytes. Using third-generation, 4-plasmid lentiviral vectors expressing EGFP, Carlotti and colleagues successfully transduced preadipocytes and mature adipocytes. Notably, adipogenesis and insulin-mediated glucose uptake in mature cells were not affected [10].

Research on adipose tissue differentiation and genetic expression has been also developed using lentiviral-based gene therapy. Crucial proteins for the functionality of adipose tissue, such as the peroxisome proliferator-activated receptor γ (PPARγ), have been silenced using lentiviral-mediated transduction of short hairpin RNA systems (shRNA) in preadipocytes. This model demonstrated that PPARγ is necessary for adipogenesis [11].

More recent in vitro work has been carried out focusing on human adipose-derived stem cells (hADSC). In order to develop new strategies for bone regeneration, the feasibility of lentiviral vectors encoding bone morphogenic protein 2 (BMP2) and NEL-like molecule 1 (NELL-1) to infect and transduce hADSC has been explored. Using a lentiviral-based vector that could be traced with EGFP, Liu and collaborators successfully transduced hADSC with BMP2/EGFP and NELL-1/EGFP. Compared to NELL-1 infected cells, BMP2 infected cells exhibited more proliferation. Moreover, 8 weeks following infection, hADSC cells continued to express NELL-1, proving the long-term transduction efficiency of the vector. Interestingly, NELL-1 expressing cells had evidence of osteogenic differentiation (increased alkaline phosphatase activation and cellular mineralization), but presented down regulation of adipogenic markers [12]. This provides another example in which lentiviral vectors transduce cells with stable gene expression over a long period of time. Furthermore, the genetic changes are resistant to cell differentiation.

To the best of our knowledge, studies of lentiviral-based vectors for gene therapy of adipose tissue in vivo are lacking. Future investigations should focus on in vivo models of disease (obesity or diabetes) and the utility of these types of vectors in this context.

1.4 Alpha and Gamma Retroviruses and Adipose Tissue

The viral family *Retroviridae* also contains the subfamilies of alpha- and gamma-retroviruses. Their structure, life cycle, and main proteins are very similar to those of lentiviruses; however, they lack accessory proteins (Vfu, Vip) and more complex transcriptional units. As stated previously, these viruses do not possess the ability to transduce nondividing cells as lentiviruses do. Nevertheless, they integrate stably into the host genome of dividing cells and

Table 2
Retroviral-based vectors and adipose tissue

	Uses	Cell type	Gene of interest	References
Lentiviral-based vector	In vitro	Human Adipose-Derived Stem Cells (ADSC)	EGFP, BMP2, NELL-1	[12]
	In vitro	Mature adipocytes	EGFP	[10]
	In vitro	Preadipocytes	PPARγ	[11]
Other retroviral-based vectors	In vitro	Preadipocytes	ACLP, PPARγ	[13, 14]
	In vitro	Human and murine ADSCs	EGFP	[9]

have the capacity of long-term transduction, which makes them good models for genome editing vectors [6].

Retroviral infection and transduction of adipose cell lines have been achieved in various in vitro models (Table 2). Retroviral vectors with forced expression of PPARγ have proven that PPARγ is sufficient to induce adipogenesis in mouse embryonic fibroblasts (MEFs) [13]. In a recent study, Gusinjac and colleagues infected 3T3-L1 cells with retroviral vectors to survey the effects of aortic carboxypeptidase-like protein (ACLP) on preadipocytes. Using a MLV-based vector (pLXSN), they effectively transduced and generated ACLP-overexpressing 3T3-L1 preadipocytes. Importantly, the cells maintained ACLP expression after differentiation into mature adipocytes [14]. When cultured in a collagen-I rich environment, ACLP-transduced cells exhibited inhibition of fatty acid synthase and PPAR-γ expression, suggesting that it may inhibit adipogenesis.

In vivo studies of retroviral vectors in adipose tissue are lacking. Though the retroviral model has been important in elucidating important aspects of adipose tissue biology, its limitation of transducing only replicating cells places a theoretical cap on the number of cells that can be successfully transduced.

1.5 Adenoviral Vectors and Adipose Tissue

Adenoviruses (adenoviridae) are a family of viruses characterized by the lack of lipid envelope and an icosahedral protein capsid encompassing a double-stranded DNA genome of approximately 36,000 base pairs. Adenovirus can infect human and animal species. Human adenoviruses have at least 57 serotypes comprising seven species, A to G. In humans, species C serotypes (1,2,5,6) are the most common. Most adenoviral-based vectors are a modified version of Ad5 (group C). The initial strategy to transition between the wild-type virus and the adenoviral vector was to delete the E1 genomic regions, which are responsible for the subsequent activation of other early-expressed genes and due to their capability of deregulating the cell cycle, possess oncogenic potential. Thus, this first generation of adenoviral vectors is replication-defective (RD) and it replaces the E1A and E1B genes with high activity promoters such as the cytomegalovirus immediate early promoter.

In general, the major drawback with adenoviral vectors lies in their strong immunogenic nature. The major capsid proteins of the virus (hexon, penton base, and fiber) generate potent innate immunologic responses and specific antibodies. After systemic administration of adenoviral vectors, a large proportion of the virus is sequestered in the liver, where they interact with resident immune cells (Kupffer cells), resulting in a massive release of pro-inflammatory cytokines (IL-6, TNF-α, IL-1β). This inflammatory response in target and surrounding tissues results in toxicity to the host and loss of the transduced cells. This is why new generations of adenoviral vectors and future directions in this field focus on reducing the host's immune response without losing transduction efficiency [15].

Adenoviral vectors are well suited for systemic delivery of adipokines in metabolic disease models. In one of the earliest studies, investigators used adenoviral vectors to replenish leptin in an *ob/ob* mouse model, achieving correction of the diabetic phenotype [16]. Lee and colleagues describe the utilization of adiponectin-expressing adenoviral vector (Ad-APN) in the *db/db* mouse model of DM2. Compared to controls, Ad-APN infected mice exhibited improvement of endothelial-dependent vasodilation, and this improvement was inhibited when administering a nitric oxide synthase inhibitor [17]. Recently, we utilized adenoviral vectors to replete the adipokine adipsin in the *db/db* mouse model. Serum expression of adipsin in Ad-adipsin mice was robustly upregulated and reached levels comparable to those of WT mice. Fasting glucose levels were significantly decreased in response to a robust increase in insulin concentrations. Interestingly, adipsin-treated *db/db* mice exhibited a significant reduction in gluconeogenic gene expression [18]. Adenoviral-mediated adipsin treatment of diabetic mice has exhibited promising results; however, the short half-life of the virus makes it a less likely candidate for human gene therapy, as diabetes is a chronic and progressive metabolic disease.

Adenoviral vectors have also been used to assess the role of specific genes and hormones in the regulation of thermogenesis and differentiation of brown and beige fat. Spiegelman and colleagues have injected adenoviral vectors expressing FNDC5, a precursor to the hormone irisin, into wild-type BALB/c mice. Plasma levels of irisin increased 3–4 fold, along with a significant upregulation of uncoupling protein 1 (UCP-1) in subcutaneous fat depots 10 days after injection. These findings suggest an important role of irisin and its precursor in regulating a thermogenic gene program in white adipose tissues. Conversely, C57BL/6 mice on a HFD were treated with FDNC5-expressing adenovirus, resulting in augmented energy expenditure and improved glucose tolerance [19]. Adenoviral vectors were used to elucidate the biological functions of Meteorin-like protein (Metrnl) on the induction of beige fat through the regulation of anti-inflammatory cytokines. Injection of adenovirus expressing metrnl resulted in a 5–6 fold increase in circulating metrnl. Interestingly, it also augmented the brown/beige fat thermogenic and mitochondrial gene program in the subcutaneous and visceral white adipose tissues, including UCP-1. These changes in gene expression were observed 5–7 days postinjection, and they returned to baseline by day 10, correlating with adenoviral expression of Metrnl. Importantly, whole-body energy expenditure was increased and glucose tolerance was improved in Metrnl-treated mice [20]. Recently, a novel fragment of Slit2 was identified as a PRDM16-regulated secreted factor from beige adipocytes and found to be important in brown/beige fat biology. Slit2 is a member of the Slit extracellular protein family. Adenoviral-based vectors expressing the C- terminal fragment of Slit-2 (Slit2-C) were used to transduce primary inguinal and brown fat adipocytes, leading to increased thermogenic gene expression (UCP-1 mRNA) at day 7. In vivo, Ad-Slit2-C augmented energy expenditure and ameliorated glucose tolerance in HFD-fed mice [21]. As one can see from these studies, adenoviral-based vectors can be powerful tools to study both in vitro and in vivo mechanisms and functions of brown and beige adipose tissue biology.

1.6 Introduction to Adeno-Associated Virus

Adeno-associated viruses (AAV) are single-strand DNA viruses belonging to the *parvoviridae* family, genus *dependoparvovirus*. They possess a genome of approximately 4.7 kb packaged in a non-enveloped icosahedral capsid. A notable characteristic of AAV is that it requires a helper virus, in most cases adenovirus or herpes virus, to efficiently infect mammalian cells. However, in the absence of a helper virus, AAVs (serotype 2) can integrate into a specific site of the mammalian genome and establish latency. To date, 12 human serotypes of AVV and more than 100 serotypes from non-human primates have been discovered. These viruses are appealing candidates for gene therapy since they are not pathogenic to humans, have a persistent expression over time, and include many serotypes with tropism for various tissues and cell types.

The main model for the first generation of AAV-based vectors was AAV-2. The initial AAV-based vectors were designed to be replication deficient. The main drawback with this approach is the limited packaging capacity (~5 kb) permitted by these viruses. It was recognized that AAV-2 had tropism for certain tissues, but could not transduce many cell types, due to their specificity for the heparan sulfate receptor. Approaches to solve these problems have led to the development of new technologies and the design of new generations of AAV-based vectors for gene delivery [22].

In order to increase the genome-carrying capacity of AAV-based vectors, trans-splicing of the AAV-vector is one of the most promising strategies. The basis for this approach is to split the transgene vector into two separate cassettes, and take advantage of the virus capacity of forming concameters through recombination in the ITRs. Other advances in the area of AAV-based vectors have helped to identify the cell surface receptors for other serotypes of the virus. For instance, AAV-3 transduces cells through heparan sulfate, AAV-4 through O-linked sialic acid, and AAV-5 through the platelet-derived growth factor receptor. Also, a specific laminin receptor has been shown to function as a receptor for AAV-2, 3, 8, and 9. Thus, a better understanding of the biology of the AAV capsid, its structure, and constitution has been fundamental for the design of recombinant AAV particles and acquiring the desired tropism. Examples of the natural tropism of specific serotypes are presented in Table 3, along with the tropism of recombinant AAV particles.

1.7 AAV Vectors and Adipose Tissue

Regarding specific transduction of adipose tissue using AAV-based vectors, Jimenez and collaborators assessed the transduction efficiency of various serotypes of AAV-based vectors (1–2 and 4–9) in white adipose tissue (WAT) and brown adipose tissue (BAT). When injecting these vectors expressing an eGFP marker protein directly into epididymal WAT (eWAT) and inguinal subcutaneous WAT (iWAT), they observed that both AAV-8 and AAV-9 serotypes outperformed other serotypes in terms of transduction efficiency. In order to achieve adipocyte-specific transduction, the adipocyte Protein 2 (aP2) promoter was used and eGFP expression was not detected in the liver or the heart. Interscapular injection of AAV-7, AAV-8, and AAV-9 vectors expressing eGFP driven by the *ucp1* promoter resulted in efficient and specific transduction of brown adipocytes. Intravascular injection of AAV-8 and AAV-9 vectors utilizing the ucp1 promoters resulted in transduction of WAT and BAT in lean and obese/diabetic *ob/ob* and *db/db* mice

Table 3
Adenoviral-based vectors and adipose tissue

	Uses	Model	Proteins expressed	References
Adenoviral—based vectors	In vivo	Delivery of adipokines/specific brown adipose tissue factors	Leptin, adiponectin, adipsin, fndc5/irisin, Metrnl, Slit 2	[16–21]
	In vitro	Transduction of primary adipocytes (white and brown)	Slit 2, Fndc5/irisin	[21]

[23]. The results from this study exemplify another great advantage of AAV-based vectors: tissue transduction with various serotypes in conjunction with the availability of strong promoters to achieve further tissue specificity.

Engineering of AAV-vectors has resulted in exciting results for adipose tissue engineering. O'Neill and colleagues generated a recombinant AAV 2/8 vector containing adiponectin promoter and microRNA-122 to achieve enhanced selectivity for adipose tissue and eliminate liver transduction by the vector. Using this vector model, they successfully transduced leptin in the SubQ WAT, visceral WAT, and BAT of *ob/ob* mice after systemic administration. Importantly, after a follow-up of 8 weeks, leptin plasma levels were significantly increased, concomitant with a reversal in weight gain, improved glucose tolerance, and decrease in hyperinsulinemia [24]. In other work, Uhrig-Schimidt and colleagues corroborated previous results using a recombinant AAV-8 vector expressing the adipose-specific promoter aP2 and successfully transduced subcutaneous and visceral fat depots in C57BL/6 mice. Through systemic administration of this vector expressing Perilipin A (PlinA), free fatty acids, glucose concentrations, and respiratory exchange ratio of these mice were improved up to 3 weeks after AAV injection [25]. Recently, Huang and colleagues assessed brown adipose tissue transduction efficiency using naturally occurring AAV phenotypes (AAV serotypes 1–8) and other engineered recombinant phenotypes (Rec1, Rec3, and Rec4). Using a different delivery approach compared to the above studies, they achieved systemic expression through oral administration of the virus. Interestingly, they found that Rec 2 exhibited greater transduction efficiency in the BAT compared to the other natural AAV serotypes and recombinant particles. Furthermore, when they compared oral administration against direct BAT injection, the authors observed similar levels of eGFP expression. As a proof of concept, the authors designed an AAV Rec 2 vector expressing vascular endothelial growth factor (VEGF) and observed overexpression of this protein in BAT, increases in brown fat mass and thermogenesis [26]. It is interesting to note from the previous studies that there are naturally and recombinant AAV vectors that have increased tropism for adipose tissue and, ever more interestingly, specificity for particular types of adipose tissue (Fig. 1).

Our own experience with AAV-vectors has focused on using this virus model as a vehicle for replenishment of the adipokine adipsin in chronic rodent models of type 2 diabetes. We have designed AAV-8 vectors expressing adipsin and injected them into *db/db* mice. Interestingly, we found that adipsin levels increase robustly in the serum to reach levels comparable to that of lean WT mice. Expression of this protein is maintained 3 months after systemic AAV administration. Our unpublished data suggest that the liver and adipose are the main tissues that express the protein.

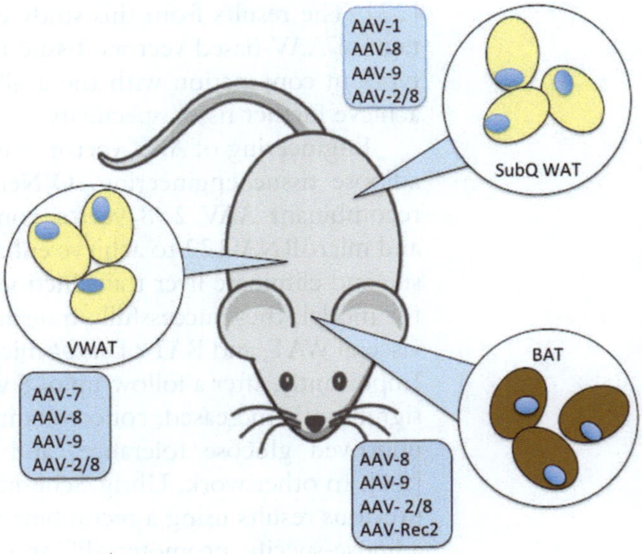

Fig. 1 Adeno-associated viral vectors for transduction of white and brown adipose tissue depots. Different types of adipose tissue such as subcutaneous white adipose tissue (SubQ WAT), visceral white adipose tissue (VWAT) and brown adipose tissue have been transduced by specific AAV serotypes (shown in the *blue box*) in vivo

2 Materials

2.1 Retrovirus and Lentivirus

1. polybrene solution at 16 mg/mL.
2. retrovirus/lentivirus vectors including transfer, packaging, and envelope vectors.
3. 293T cells.
4. adipocyte cultures.
5. transfection reagents.
6. luer lock syringe and 0.45 micron syringe filter.

2.2 Adenovirus

1. adenovirus vector.
2. 293A cells.
3. adipocyte cultures.
4. adenovirus purification system.
5. adenovirus titering system.
6. transfection reagents.

2.3 Adeno-Associated Virus

1. AAV vector.
2. 293T cells.
3. AAV purification method.
4. AAV titering system.

5. adipocyte culture.
6. transfection reagents.

3 Methods

3.1 Production of Retro/Lentiviral Vectors to Transduce Adipocyte Cultures

1. Transfect the retro/lentiviral vectors into 293T cells via a transient transfection method of choice (*see* **Notes 1** and **5**).
2. Collect the viral particle containing supernatant during peak expression (typically 48–72 h posttransfection). Filter supernatant through a 0.45 micron syringe filter. Store viral particles (filtered supernatant) at 4 °C prior to use or at −80 °C if not used within 24 h.
3. Apply viral particles containing polybrene (1–8 μg/mL) at a ratio of 1:1 to 1:10 to cultured adipocytes[1] for 12–16 h. We recommend that the user titrate the amount of polybrene and viral particles to achieve the desired effect.
4. Change media after overnight incubation.
5. Refresh media daily or every other day until ready for experiment.
6. Assess for effects on gene expression, phenotype, and function 2–10 days following viral transduction.
7. Assess for effects on adipocyte differentiation using either Oil Red O staining or RT-qPCR for adipocyte-specific genes such as *adipsin* and *pparg2* (*see* **Note 6**).

3.2 Production of Adenoviral Vectors to Transduce Adipocyte Cultures

1. Transfect the adenoviral vectors into 293A cells via a transient transfection method of choice (*see* **Notes 1** and **5**).
2. Harvesting adenovirus. Scrape and collect the cells along with the supernatant when they show 50–80 % cytopathic effect (typically 1 week later). Apply three rounds of freeze/thaw cycles to collected cells to lyse the cells and release adenoviral particles into the supernatant. After third thaw, spin cells at 3000 × *g* for 15 min at 4 °C to pellet the cellular debris. Transfer the supernatant into a new tube. This supernatant is now the crude viral extract. The crude viral extract can be amplified by transducing new and larger batches of 293A cells (scalable to dozens of 15 cm tissue culture plates). Aliquot and store the crude viral extract at −80 °C. When starting out with

[1] Viral particles can be added to cultured cells during the preadipocyte stage to test for effects on adipogenesis or to differentiated adipocytes to assess for effects on gene expression and adipocyte function. For retrovirus, viral particles will need to be added during the preadipocyte stage when the cells are actively undergoing replication

a 10 cm tissue culture plate, we recommend making 1 mL aliquots (*see* **Note 3**).

3. Optional. Adenoviral particles can be purified and concentrated by various protocols including cesium chloride density gradient or using a chromatography approach with membranes that selectively bind the adenovirus. Store aliquots at −80 °C. This purification step is recommended for in vivo use.
4. Titering of adenovirus can be performed with antibody-based detection assays against viral proteins or with markers of gene expression such as GFP that are present on some adenoviral vectors. This is typically carried out by making 10-fold serial dilutions from 10^{-2} to 10^{-7} of the adenovirus and incubating it with 293A cells in a 12-well plate. 48 h later, cells that have been infected are counted based on expression of viral proteins or fluorescence of the GFP marker protein (*see* **Note 4**).
5. Apply adenovirus to cultured adipocytes and incubate for 12–16 h prior to refreshing the media. The optimal dose of adenovirus varies depending on the application. We suggest titrating the dose of the virus for the user's experiment. Typical doses for one well of a 12-well plate range from 10^7 to 10^9 IFU.
6. Change the media the following day.
7. Change the media daily or every other day until ready for harvest or experiment.
8. Assess for effects on gene expression, phenotype, and function 2–3 days following viral transduction.

3.3 Production of AAV Vectors to Transduce Adipocyte Cultures

1. Transfect the AAV vectors into 293T cells via a transient transfection method of choice (*see* **Notes 1** and **5**).
2. Harvesting AAV. Scrape and collect the cells 72 h after transfection. Pellet cells at 2000 g for 15 min at 4 °C and remove the supernatant.
3. The AAV cell pellet can now be purified and concentrated by standard protocols including cesium chloride density gradient and iodixanol ultracentrifugation or using a chromatography approach with membranes that selectively bind the AAV serotype.
4. Titering of AAV can be performed with antibody-based detection assays against viral proteins or with markers of gene expression such as GFP that are present on some AAV vectors. This is typically carried out by making tenfold serial dilutions from 10^{-2} to 10^{-7} of the adenovirus and incubating it with 293A cells in a 12-well plate. 48 h later, cells that have been infected are counted based on expression of viral proteins or fluorescence of the GFP marker protein (*see* **Note 4**).

5. Apply AAV to cultured adipocytes and incubate for 12–16 h prior to refreshing the media. The optimal dose of AAV varies depending on the application. We suggest titrating the dose of the virus for the user's experiment. Typical doses for one well of a 12-well plate range from 10^7 to 10^9 IFU.
6. Change media the following day.
7. Refresh media daily or every other day until cells are ready for harvest or experiment.
8. Assess for effects on gene expression, phenotype, and function 2–7 days following viral transduction.
9. Assess for effects on adipocyte differentiation using either Oil Red O staining or RT-qPCR for adipocyte-specific genes such as *adipsin* and *pparg2* (*see* **Note 6**).

3.4 Adenovirus and AAV Vectors for In Vivo Transduction of Mice

1. Prior to adenovirus or AAV injection, thaw the required amount of virus on ice (*see* **Note 2**).
2. Dilute the adenovirus or AAV to the desired concentration with ice cold normal saline. Mix gently and do not vortex. Depending on the target tissue or site of administration, the volume injected should be adjusted. Typical injection volumes range from 20 to 200 μL. The amount of virus injected should also be titrated depending on the target tissue and the anticipated phenotype. The doses of the virus range from 10^8 to 10^{11} IFU.
3. Keep the diluted virus on ice until just prior to injection. A few minutes prior to injection, allow the virus to warm to room temperature.
4. Inject the virus into the mouse or experimental animal using a syringe needle or equivalent.
5. Assess for effects on gene expression or physiology. For adenovirus, we recommend a period of 3–7 days following infection. For AAV, analysis can be performed starting 3 days to months after infection.

4 Notes

1. Biohazard safety conditions should be followed stringently for each of the viruses and cell culture with the biosafety level (BL) guidelines appropriate to the type of virus. The NIH has assigned AAV a BL1 containment for laboratory procedures. For retroviruses, BL2 or enhanced BL2 containment is recommended in the laboratory setting. The NIH has assigned adenovirus BL2 containment for laboratory procedures. This includes attention to sharps, use of safety needles, and the use of personal protective equipment safety (eye goggles, lab coat,

skin gloves) to reduce potential mucosal contact with the vector [27]. Consult with your institution's laboratory safety office before the use of viral vectors in the lab.

2. Adenovirus and AAV are labile and lose their titer upon heating and with freeze/thaw cycles. We therefore recommend thawing the adenovirus and AAV on ice prior to use. We suggest aliquoting the adenovirus prior to long term storage at −80 °C.
3. Amplification of adenovirus induces more rapid cytotoxicity than initial transient transfection. Be prepared to harvest the adenovirus 2–3 days after infection from crude extract with good quality preps.
4. Viral titers depend on cell condition, vector packaging generation, and the design of the transfer vector. As more plasmids are used to deliver the necessary genes, the efficiency of generating productive viral particles is diminished. Thus, third generation four construct vectors generate lower titers than three construct ones [8].
5. Typical transient transfection methods used in our laboratory include calcium phosphate and lipid-based methods such as lipofectamine (Invitrogen).
6. Viral transduction can have effects on adipocyte differentiation in vitro. The transduced genes of interest may have additional effects on adipocyte differentiation. The potential effects on adipocyte differentiation may or may not be what the investigator is studying but should be evaluated. Assessment of adipocyte differentiation can be done by Oil Red O staining or RT-qPCR for adipocyte-specific genes such as *adipsin* and *pparg2*.

4.1 Conclusion

The study of adipose tissue biology and the role of adipose-derived factors in metabolic diseases is a growing area of research. Viral vectors have become crucial tools for genome editing in this context. Retroviral-based vectors, including lentiviral vectors, have been used primarily in the in vitro context for transduction of adipose-derived stem cells, preadipocytes, and mature adipocytes from various animal models. Consequently, important discoveries in the area of adipose tissue development and the biology of the transcriptional and regulatory proteins crucial in adipogenesis have been developed with the aid of retroviral vectors. On the other hand, adenoviral vectors have been mainly utilized as delivery vehicles for hormones and adipokines in metabolic disease models. Adenoviral vectors offer a simple mechanism to achieve systemic expression with high transduction efficiency. However, in chronic disease models, short-term expression is a major drawback. Currently, retrovirus and adenovirus do not offer much of an advantage for adipose specificity when used in vivo.

Adeno-associated viral vectors offer long-term protein expression and have been actively used as adipokine vehicles in metabolic disease models. The variety of AAV serotypes and the development of different recombinant particles through the engineering of capsid particles enable wide tropism and the possibility of achieving adipose-specific transduction. This is important to distinguish biological mechanisms and secreted factors between the different types of adipose tissues (brown, white, beige, visceral, subcutaneous). In the future, with the advent of more adipose-specific and even adipose depot-specific vectors one can hope to target specific adipose depots.

In this section, we have mostly provided examples with gain of function approaches. On the flip side, it is possible to engineer viral vectors for loss of function experiments. For example, a BAT-specific AAV expressing Cre may be combined with mice with a floxed gene of interest to generate BATspecific knockout mice of the gene of interest. In addition, shRNAs and mutant proteins with dominant negative phenotypes can be cloned into a viral vector of choice.

Given the recent advances and increasing recognition of adipose tissue as an endocrine organ with focal roles in metabolic diseases such as obesity and diabetes, future efforts utilizing viral vectors for the delivery of adipokines and transduction of adipose tissue in human clinical trials may be a promising path to take in the fight against cardiometabolic diseases.

References

1. Rosen ED, Spiegelman BM (2014) What we talk about when we talk about fat. Cell 156(1–2):20–44. doi:10.1016/j.cell.2013.12.012
2. Baker KD, Thummel CS (2007) Diabetic larvae and obese flies-emerging studies of metabolism in Drosophila. Cell Metab 6(4):257–266. doi:10.1016/j.cmet.2007.09.002
3. Hotamisligil GS (2006) Inflammation and metabolic disorders. Nature 444(7121):860–867. doi:10.1038/nature05485
4. Lelliott C, Vidal-Puig AJ (2004) Lipotoxicity, an imbalance between lipogenesis de novo and fatty acid oxidation. Int J Obes Relat Metab Disord 28(Suppl 4):S22–S28. doi:10.1038/sj.ijo.0802854
5. Zechner R, Kienesberger PC, Haemmerle G, Zimmermann R, Lass A (2009) Adipose triglyceride lipase and the lipolytic catabolism of cellular fat stores. J Lipid Res 50(1):3–21. doi:10.1194/jlr.R800031-JLR200
6. Maier P, von Kalle C, Laufs S (2010) Retroviral vectors for gene therapy. Future Microbiol 5(10):1507–1523. doi:10.2217/fmb.10.100
7. Gilbert JR, Wong-Staal F (2001) HIV-2 and SIV vector systems. Somat Cell Mol Genet 26(1–6):83–98
8. Sakuma T, Barry MA, Ikeda Y (2012) Lentiviral vectors: basic to translational. Biochem J 443(3):603–618. doi:10.1042/BJ20120146
9. Morizono K, De Ugarte DA, Zhu M, Zuk P, Elbarbary A, Ashjian P, Benhaim P, Chen IS, Hedrick MH (2003) Multilineage cells from adipose tissue as gene delivery vehicles. Hum Gene Ther 14(1):59–66. doi:10.1089/10430340360464714
10. Carlotti F, Bazuine M, Kekarainen T, Seppen J, Pognonec P, Maassen JA, Hoeben RC (2004) Lentiviral vectors efficiently transduce quiescent mature 3T3-L1 adipocytes. Mol Ther 9(2):209–217. doi:10.1016/j.ymthe.2003.11.021
11. Katayama K, Wada K, Miyoshi H, Ohashi K, Tachibana M, Furuki R, Mizuguchi H, Hayakawa T, Nakajima A, Kadowaki T, Tsutsumi Y, Nakagawa S, Kamisaki Y,

Mayumi T (2004) RNA interfering approach for clarifying the PPARγ pathway using lentiviral vector expressing short hairpin RNA. FEBS Lett 560(1–3):178–182. doi:10.1016/s0014-5793(04)00100-0

12. Liu Y, Chen C, He H, Wang D, E L, Liu Z, Liu H (2012) Lentiviral-mediated gene transfer into human adipose-derived stem cells: role of NELL1 versus BMP2 in osteogenesis and adipogenesis in vitro. Acta Biochim Biophys Sin (Shanghai) 44(10):856–865. doi:10.1093/abbs/gms070
13. Tontonoz P, Hu E, Spiegelman BM (1994) Stimulation of adipogenesis in fibroblasts by PPAR gamma 2, a lipid-activated transcription factor. Cell 79(7):1147–1156
14. Gusinjac A, Gagnon A, Sorisky A (2011) Effect of collagen I and aortic carboxypeptidase-like protein on 3T3-L1 adipocyte differentiation. Metabolism 60(6):782–788. doi:10.1016/j.metabol.2010.07.028
15. Wold WS, Toth K (2013) Adenovirus vectors for gene therapy, vaccination and cancer gene therapy. Curr Gene Ther 13(6):421–433
16. Muzzin P, Eisensmith RC, Copeland KC, Woo SL (1996) Correction of obesity and diabetes in genetically obese mice by leptin gene therapy. Proc Natl Acad Sci U S A 93(25):14804–14808
17. Lee S, Zhang H, Chen J, Dellsperger KC, Hill MA, Zhang C (2012) Adiponectin abates diabetes-induced endothelial dysfunction by suppressing oxidative stress, adhesion molecules, and inflammation in type 2 diabetic mice. Am J Physiol Heart Circ Physiol 303(1):H106–H115. doi:10.1152/ajpheart.00110.2012
18. Lo JC, Ljubicic S, Leibiger B, Kern M, Leibiger IB, Moede T, Kelly ME, Chatterjee Bhowmick D, Murano I, Cohen P, Banks AS, Khandekar MJ, Dietrich A, Flier JS, Cinti S, Bluher M, Danial NN, Berggren PO, Spiegelman BM (2014) Adipsin is an adipokine that improves beta cell function in diabetes. Cell 158(1):41–53. doi:10.1016/j.cell.2014.06.005
19. Bostrom P, Wu J, Jedrychowski MP, Korde A, Ye L, Lo JC, Rasbach KA, Bostrom EA, Choi JH, Long JZ, Kajimura S, Zingaretti MC, Vind BF, Tu H, Cinti S, Hojlund K, Gygi SP, Spiegelman BM (2012) A PGC1-alpha-dependent myokine that drives brown-fat-like development of white fat and thermogenesis. Nature 481(7382):463–468. doi:10.1038/nature10777
20. Rao RR, Long JZ, White JP, Svensson KJ, Lou J, Lokurkar I, Jedrychowski MP, Ruas JL, Wrann CD, Lo JC, Camera DM, Lachey J, Gygi S, Seehra J, Hawley JA, Spiegelman BM (2014) Meteorin-like is a hormone that regulates immune-adipose interactions to increase beige fat thermogenesis. Cell 157(6):1279–1291. doi:10.1016/j.cell.2014.03.065
21. Svensson KJ, Long JZ, Jedrychowski MP, Cohen P, Lo JC, Serag S, Kir S, Shinoda K, Tartaglia JA, Rao RR, Chedotal A, Kajimura S, Gygi SP, Spiegelman BM (2016) A secreted Slit2 fragment regulates adipose tissue thermogenesis and metabolic function. Cell Metab 23(3):454–466. doi:10.1016/j.cmet.2016.01.008
22. Daya S, Berns KI (2008) Gene therapy using adeno-associated virus vectors. Clin Microbiol Rev 21(4):583–593. doi:10.1128/CMR.00008-08
23. Jimenez V, Munoz S, Casana E, Mallol C, Elias I, Jambrina C, Ribera A, Ferre T, Franckhauser S, Bosch F (2013) In vivo adeno-associated viral vector-mediated genetic engineering of white and brown adipose tissue in adult mice. Diabetes 62(12):4012–4022. doi:10.2337/db13-0311
24. O'Neill SM, Hinkle C, Chen SJ, Sandhu A, Hovhannisyan R, Stephan S, Lagor WR, Ahima RS, Johnston JC, Reilly MP (2014) Targeting adipose tissue via systemic gene therapy. Gene Ther 21(7):653–661. doi:10.1038/gt.2014.38
25. Uhrig-Schmidt S, Geiger M, Luippold G, Birk G, Mennerich D, Neubauer H, Grimm D, Wolfrum C, Kreuz S (2014) Gene delivery to adipose tissue using transcriptionally targeted rAAV8 vectors. PLoS One 9(12):e116288. doi:10.1371/journal.pone.0116288
26. Huang W, McMurphy T, Liu X, Wang C, Cao L (2016) Genetic manipulation of brown fat via oral administration of an engineered recombinant adeno-associated viral serotype vector. Mol Ther 24(6):1062–1069. doi:10.1038/mt.2016.34
27. Durand S, Cimarelli A (2011) The inside out of lentiviral vectors. Viruses 3(2):132–159. doi:10.3390/v3020132

Chapter 12

Bioenergetic Analyses in Adipose Tissue

Lawrence Kazak

Abstract

Adipose tissue plays an important role in whole body metabolism, and its dysfunction results in poor metabolic health and is associated with a diverse set of disorders, such as obesity, cardiovascular disease, hypertension, cancer, and type 2 diabetes. The search for cellular processes that can be targeted to activate adipose oxidative metabolism is thought to hold promise for combating these diseases, and so the analysis of adipose bioenergetics is crucial toward this end. A description is provided here for the preparation of mitochondria and primary adipocytes from adipose tissue to be used for bioenergetics profiling.

Key words Mitochondria, Brown fat, Beige fat, Bioenergetics, Primary adipocytes, Stromal vascular fraction

1 Introduction

Bioenergetics can be examined using isolated mitochondria and intact cells. While a Clark-type oxygen electrode has been used to measure mitochondrial respiration for over half a century [1], recently developed plate-based assays (XF Analyzer) can provide the same information [2, 3].

The Clark-type electrode is limited to cells that can be obtained in high yield and in suspension. The Seahorse XF24 and XF96 extracellular flux analyzer is rapidly becoming the tool of choice for many investigators because of some key advantages. Given that mitochondrial viability is time sensitive following isolation, the ability to analyze multiple conditions simultaneously provides an important benefit. In addition, the chamber volume size is as small as 7 μL, and so the amount of material needed can be scaled down quite substantially. This is particularly important when investigators are interested in examining mitochondrial function of cells and tissues that have either low mitochondrial abundance, or the tissue of interest is difficult to come by, such as may be the case for human clinical samples. Additional reading on the advantages of various approaches to monitoring mitochondrial bioenergetics can be found in the following reviews [2, 3].

Jun Wu (ed.), *Thermogenic Fat: Methods and Protocols*, Methods in Molecular Biology, vol. 1566,
DOI 10.1007/978-1-4939-6820-6_12, © Springer Science+Business Media LLC 2017

Mitochondria obtained by differential centrifugation alone have a certain amount of contaminating organelles, and so should be thought of as enrichments. The extent of contamination is also tissue-dependent. A further enrichment of mitochondria (removal of contaminating organelles and cellular structures) can be obtained by ultracentrifugation of mitochondria through a sucrose gradient [4].

Bioenergetic function in intact cells can be monitored with comparable accuracy to studies with isolated mitochondria. Some disadvantages to working with intact cells are that the mitochondria cannot be accessed directly by the full range of substrates and inhibitors typically used in studies using purified organelles. However, the benefit of using cultured cells is that mitochondrial function can be examined in a physiologically relevant environment. Direct comparison of incubation conditions of primary adipocytes has demonstrated that primary adipocyte differentiation is more efficient at 10 % CO_2 compared to 5 % CO_2.

2 Materials

2.1 Mitochondrial Isolation for Functional Analysis

1. Homogenizer with tight-fitting teflon pestle.
2. Polyallomer Centrifuge tubes.
3. Malic acid: Dissolve to a stock concentration of 0.5 M, pH to 7.2 with KOH (aliquot and store at −20 °C). Use at 5 mM.
4. Sodium Pyruvate: Dissolve in respiration buffer to a stock concentration of 0.5 M. Use fresh at 5 mM.
5. Potassium Succinate anhydrous: Dissolve to a stock concentration of 0.5 M, pH to 7.2 with KOH (aliquot and store at −20 °C). Use at 5 mM.
6. Glycerol 3-phosphate: Dissolve to a stock concentration of 0.5 M, pH to 7.2 with KOH (aliquot and store at −20 °C). Use at 5 mM.
7. GDP: Dissolve to a stock concentration of 0.5 M, pH to 7.2 with KOH (aliquot and store at −20 °C). Use at 1 mM.
8. ADP: Dissolve to a stock concentration of 0.5 M, pH to 7.2 with KOH (aliquot and store at −20 °C). Use at 0.1–1 mM.
9. Oligomycin: Dissolve in 96 % ethanol to a stock concentration of 5 mM. Aliquot and store at −20 °C. Use at 14 μM.
10. FCCP: Dissolve in DMSO to a stock concentration of 5 mM. Aliquot and store at −20 °C. Use at 10 μM.
11. Antimycin A: Dissolve in 96 % ethanol to a stock concentration of 5 mM. Aliquot and store at −20 °C. Use at 4 μM.
12. Rotenone: Dissolve in DMSO to a stock concentration of 5 mM. Aliquot and store at −20 °C. Use at 4 μM.
13. 20 % BSA: Dissolve 20 g BSA in 100 mL distilled water and store at −20 °C.

14. XF Analyzer Cartridges + Plates (Seahorse Bioscience).
15. SHE buffer (store at 4 °C—use within 2 weeks): 250 mM Sucrose; 5 mM Hepes, pH 7.4; 1 mM EGTA.
16. SHEB buffer (use fresh): SHE buffer + 2 % BSA.
17. Storage buffer: 100 mM KCl; 20 mM TES, pH 7.2; 1 mM EDTA (*see* **Note 1**).
18. Sucrose respiration buffer, pH 7.2 (without substrate): 125 mM Sucrose; 20 mM TES, pH 7.2; 2 mM $MgCl_2$; 1 mM EDTA; 4 mM KH_2PO_4; 4 % BSA (*see* **Note 2**).
19. KCl respiration buffer, pH 7.2 (without substrate): 50 mM KCl; 4 mM KH_2PO_4; 5 mM HEPES; 1 mM EGTA; 4 % BSA (*see* **Note 2**).

2.2 Sucrose-Gradient Mitochondrial Purification

1. Polyallomer Centrifuge tubes.
2. 20 % BSA.
3. Gradient buffer (prepare fresh): 10 mM HEPES, pH 7.4, 5 mM EDTA, and 2 mM DTT.

2.3 Isolation and Storage of Stromal Vascular Fraction from Subcutaneous and Brown Adipose Tissue

1. Adipocyte Culture Medium: DMEM/F12, 10 % FBS, and 0.1 % PenStrep.
2. FBS: Use 50/500 mL DMEM.
3. Collagenase B: *Store powder at 4 °C.*
4. Collagenase D: *Store powder at 4 °C.*
5. Dispase II: *Store 50× stock (155 U/mL) aliquots at −20 °C.*
6. Cell Strainers: 100 and 40 μm.
7. SubQ Collagenase solution (10 mL solution in PBS): 10 mg/mL collagenase D, 3 U/mL dispase II, and 10 mM $CaCl_2$.
8. BAT Collagenase solution (10 mL solution in water): 1.5 mg/mL collagenase B, 123 mM NaCl, 5 mM KCl, 1.3 mM $CaCl_2$, 5 mM glucose, 100 mM HEPES, and 4 % BSA.
9. Freezing medium: 90 % FBS; 10 % DMSO.

2.4 Bioenergetic Analysis from Cultured Primary Adipocytes

1. Rosiglitazone: dissolve stock to 10 mM in DMSO, store at −20 °C.
2. Isobutylmethylxanthine (IBMX): dissolve stock to 250 mM in DMSO, store at −20 °C.
3. Dexamethasone: dissolve stock to 5 mM in absolute ethanol, store at −20 °C.
4. Insulin: dissolve stock to 870 μM in 50 mM HCl, store at −20 °C until used, then continue to store at 4 °C.
5. T3: dissolve stock to 1 mM in 1 N NaOH. Make 1:100 dilution in sterile PBS to obtain working stock of 10 μM, store at −20 °C.

6. Indomethacin: dissolve stock to 125 mM in absolute ethanol, store at −20 °C.
7. Inguinal differentiation cocktail: 1 μM rosiglitazone, 0.5 mM isobutylmethylxanthine, 1 μM dexamethasone, and 5 μg/mL insulin.
8. Inguinal maintenance cocktail: 1 μM rosiglitazone and 5 μg/mL insulin.
9. Brown differentiation cocktail: 1 μM rosiglitazone, 0.5 mM isobutylmethylxanthine, 5 μM dexamethasone, and 0.114 μg/mL insulin, 1 nM T3, and 125 μM Indomethacin.
10. Brown maintenance cocktail: 1 μM rosiglitazone 1 nM T3 rosiglitazone and 0.5 μg/mL insulin.
11. Oligomycin: Dissolve in 96 % ethanol to a stock concentration of 5 mM. Aliquot and store at −20 °C. Use at 3 μM.
12. FCCP: Dissolve in DMSO to a stock concentration of 5 mM. Aliquot and store at −20 °C. Use at 2 μM.
13. Antimycin A: Dissolve in 96 % ethanol to a stock concentration of 5 mM. Aliquot and store at −20 °C. Use at 4 μM.
14. Rotenone: Dissolve in DMSO to a stock concentration of 5 mM. Aliquot and store at −20 °C. Use at 4 μM.
15. 20 % BSA.
16. Unbuffered DMEM (1 L), store at 4 °C:
    - Bottle of DMEM powder (Sigma; D5030).
    - 1.85 g NaCl.
    - OPTIONAL: 3 mg phenol red.
    - pH to 7.4.
    - filter-sterilize.
17. Respiration media: Unbuffered DMEM, supplemented with 2 % BSA. pH to 7.2 with NaOH.
18. XF Analyzer Cartridges + Plates (Seahorse Bioscience).

3 Method

3.1 Crude Mitochondrial Isolation

Carry out entire mitochondrial isolation procedure at 4 °C. Keep mitochondria concentrated and be gentle during pipetting. Mitochondria from interscapular brown adipose tissue (BAT) can be obtained easily from five mice, housed at a temperature of 22 °C. Mitochondrial yield is increased following cold exposure. One week of cold exposure (4 °C) or treatment of mice with the β3-agonist, CL-316,243 (1mg/kg), should be sufficient to obtain a sufficient amount of beige/brite mitochondria from the subcutaneous depot of 5–10 mice.

1. Animals are rapidly killed by cervical dislocation.
2. Dissect fat pads and place into ice-cold SHE buffer.
3. Mince tissues with scissors until a consistency of thick paste is reached.
4. Add minced tissue to SHEB buffer (~ 10 mL SHEB buffer per gram tissue).

OPTIONAL: Homogenize with Tissue Master to disrupt tissue further (*see* **Note 3**).

5. Transfer tissue slurry to glass homogenizer with a tight-fitting teflon pestle. Homogenize until smooth consistency and it is easy to move the Teflon pestle all the way to the bottom of the glass homogenization tube without the appearance of tissue chunks (*see* **Note 4**).
6. Filter homogenate through two layers of cheesecloth.
7. Spin 8500 × *g* for 10 min at 4 °C (*see* **Note 5**).
8. Discard fat layer by inverting tube into waste.
9. Resuspend pellet (containing nuclei, unbroken cells, and mitochondria) in 2 mL of SHEB buffer (*see* **Note 6**).
10. Volume up to 10 mL with SHEB buffer.
11. Spin 700 × *g* for 5 min at 4 °C. Transfer supernatant to clean tube. Repeat spin to pellet residual debris.
12. Transfer supernatant (containing mitochondria) to a new tube. Spin 8500 × *g* for 10 min at 4 °C to pellet mitochondria.
13. Resuspend mitochondria in a small volume of storage buffer (~ 0.2 mL). Measure protein concentration. Storing organelles at high concentrations (~ 30–50 mg/mL) increases viability and the length of time where respiratory control can be achieved.
14. Add mitochondria to the XF cell culture microplate (on ice) in a volume of 50 μL. This small volume ensures that organelles will attach to the bottom of the well, and not the sides (*see* **Note 7**).
15. Spin plate at 2000 × *g* for 20 min at 4 °C to pellet mitochondria to bottom of well.
16. While plate is spinning, add compounds to injection ports of XF cartridge (*see* **Note 8**).
17. Equilibrate cartridge.
18. After spinning mitochondria, place XF microplate on ice.
19. Add 450 μL respiration buffer to each well.

20. Incubate at 37 °C for 10 min (*see* **Note 9**).
21. Transfer mitochondria to XF Analyzer and begin respiration measurements.

XF program:

1. **State IV: Basal respiration**
   - Mix: 20 s
   - Measure 3 min
   - Mix: 20 s
   - Measure 3 min
   - Mix: 20 s (*see* **Note 10**)
2. **Inject port A:**
   - Mix: 20 s (*see* **Note 11**)
   - Measure 3 min
   - Mix: 20 s
   - Measure 3 min
   - Mix: 20 s
3. **Inject port B–D** (same as after injection into port A)

3.2 Sucrose-Gradient Mitochondrial Purification

Mitochondria obtained by differential centrifugation have a certain amount of contaminating organelles, and so should be thought of as enrichments. The extent of contamination is also tissue-dependent. A further enrichment of mitochondria (removal of contaminating organelles and cellular structures) can be obtained by ultracentrifugation of mitochondria through a sucrose gradient [4, 5, 6].

1. Isolate crude mitochondria by differential centrifugation and store concentrated on ice.
2. Dissolve sucrose to a concentration of 1 M (0.342 g/mL) and 1.5 M (0.513 g/mL) in gradient buffer.
3. Layer sucrose solutions into a polyallomer centrifuge tube.
4. Load mitochondrial samples on top of gradient.
5. Ultracentrifuge at 97,000 × *g* for 1 h at 4 °C.
6. Intact organelles band at the sucrose interface (broken organelles stay on top, and larger contaminating structures will pellet to the bottom of the tube).
7. Carefully extract the mitochondria at the interface and wash twice in SHE buffer and centrifuging 8500 × *g* for 10 min at 4 °C each time.
8. LC-MS/MS or Western blotting can be used to monitor mitochondrial enrichment.

3.3 Isolation and Storage of Stromal Vascular Fraction from Subcutaneous and Brown Adipose Tissue

1. Five mice (4–6 weeks old) are rapidly euthanized by cervical dislocation.
2. Dissect fat pads and place into PBS at room temperature.
3. Mince tissues with spring scissors for 5–7 min.
4. Digest minced tissue for 45 min at 37 °C in SubQ Collagenase solution.
5. Add 20 mL of adipocyte culture medium. Filter tissue suspension through a 100 μm cell strainer and centrifuge at 600 × *g* for 5 min to pellet the SVF.
6. Resuspended cell pellet in 10 mL adipocyte culture medium. Filter through a 40 μm cell strainer.
7. Centrifuge at 600 × *g* for 5 min.
8. Resuspend SVF pellet in 10 mL of adipocyte culture medium and incubate in 10 cm cell culture dish overnight.
9. The next morning, wash cells several times until most debris is removed. Subcutaneous preadipotyes should remain attached.
10. The dish usually becomes ~70–80 % confluent 2–3 days after plating.

3.3.1 Primary Subcutaneous Stromal Vascular Fraction (SVF) Isolation

3.3.2 Primary Brown Fat SVF Isolation

Interscapular brown adipose SVF can be obtained from 2 to 6 day old pups (*see* **Note 12**).

1. Five mice are rapidly euthanized by cervical dislocation.
2. Dissect fat pads and place into PBS at room temperature.
3. Mince tissues with spring scissors for 5–7 min.
4. Digest minced tissue for 45 min at 37 °C in BAT Collagenase solution.
5. Add 20 mL of adipocyte culture medium. Filter tissue suspension through a 100 μm cell strainer and centrifuge at 600 × *g* for 5 min to pellet the SVF.
6. Resuspended cell pellet in 10 mL adipocyte culture medium. Filter through a 40 μm cell strainer.
7. Centrifuge at 600 × g for 5 min.
8. Resuspend SVF pellet in 10 mL of adipocyte culture medium and incubate in 10 cm cell culture dish overnight.
9. The next morning, wash cells several times until most debris is removed. Brown preadipocytes should remain attached.
10. The dish usually becomes ~70–80 % confluent 2–3 days after plating.

3.3.3 Storage of Primary SVF

Subcutaneous and brown preadipocytes can be expanded (albeit to a limited degree) and stored frozen for use at a later time.

1. 2–3 days following the initial plating of cells. The plate will become ~70–80 % confluent.

2. Remove adipocyte culture medium and wash with warm PBS. Remove PBS.
3. Add 1 mL of 0.25 % Trypsin-EDTA solution. Incubate at 37 °C for 3 min.
4. Dilute cells in 30 mL media and plate into three 10 cm dishes. These dishes will become ~70–80 % confluent in 2 days.
5. Wash, trypsinize, and split the three plates of cells into ten 10 cm dishes. These dishes will become ~70–80 % confluent in 2 days.
6. Wash, trypsinize, and pellet the cells in a single tube.
7. Resuspend the pellet in 10 mL of freezing medium, and aliquot 1 mL of cell suspension into cryovials and freeze slowly at −80 °C overnight. The following day, cells can be transferred to storage in liquid nitrogen long term.

3.4 Bioenergetic Analysis from Cultured Primary Adipocytes

3.4.1 Adipocyte Differentiation (See Note 13)

1. Rapidly thaw one vial of primary preadipocytes.
2. Remove DMSO by diluting the cells in 5 mL of adipocyte culture medium and centrifuging at 1000 × *g* for 3 min at room temperature. Remove supernatant.
3. Resuspend cells in adipocyte culture medium and plate cells in 10 mL in a 10 cm dish.
4. The following day change media on cells.
5. The following evening (~2 days after the thawing), wash, trypsinize, and count cells.

3.4.2 Primary Inguinal Adipocyte Differentiation

1. In the evening, 20,000 inguinal preadipocytes are plated per well of a seahorse plate.
2. The following morning, add inguinal differentiation cocktail for 2 days.
3. Refeed cells every 2 days with inguinal maintenance cocktail.
4. Cells are ready for bioenergetic analyses by day 5 or 6 postdifferentiation.

3.4.3 Primary Brown Adipocyte Differentiation

1. In the evening, 15,000 brown preadipocytes are plated per well of a seahorse plate.
2. The following morning, add brown differentiation cocktail for 2 days.
3. Refeed cells every 2 days with brown maintenance cocktail.
4. Cells are ready for bioenergetic analyses by day 5 or 6 postdifferentiation.

3.4.4 Bioenergetic Analyses of Primary Adipocytes Using the XF Analyzer

1. Hydrate XF cartridge the night before.
2. Aspirate media from cells.
3. Wash cells with 250 μL of Unbuffered DMEM (supplemented with substrate of choice). Media should be supplemented with 2 % essentially fatty acid free BSA (*see* **Note 14**).

4. Aspirate and add 600 μL of Unbuffered DMEM to cells and incubate at 37 °C in an incubator without CO_2 (up to 1 h).
5. Prepare compounds for injection and add 75 μL of each compound into injection ports in cartridge.
6. Equilibrate cartridge (~20–30 min).
7. When cartridge equilibration is complete, remove plate with equilibration media and load plate with cells.
8. Start the run.

4 Notes

1. If mitochondria are stored on ice in 250 mM SHE buffer, they need to be incubated in hypotonic respiration buffers for ~10 min to expand the matrix for oxidation of substrates to occur [7].
2. 4 % BSA is sufficient to sequester endogenous free fatty acids [8].
3. Tissue master homogenization will increase yield obtained from subcutaneous and visceral depots.
4. Use 6–8 strokes for BAT, and up to 20 strokes for subcutaneous and visceral WAT.
5. The cellular debris and mitochondria will pellet, while the lipid will float on the supernatant.
6. It is easier to obtain a homogenous suspension using a small volume.
7. The amount of mitochondria should be determined empirically and will depend on the tissue and substrate used. 15 μg mitochondrial protein from the subcutaneous depots typically gives us an oxygen consumption rate of 100–150 pmoles O_2/min, respiring on pyruvate and malate.
8. Remember that the stock concentration of each compound must be adjusted accordingly to control for the increase in well volume that occurs following each successive addition through the injection ports.
9. Aim to have the cartridge equilibration completed before the 10 min.
10. This mixing before injection is to help sensors reequilibrate after a measurement.
11. This mixing is to mix the injected compound.
12. It may be the case that genetically modified mice require genotyping prior to pooling animals before preadipocyte isolation. If this is the case, it may be necessary to wait until the mice become older than 2–6 days to genotype. We have obtained high-quality brown preadipocytes with robust differentiation potential and response to adrenergic stimulation from these older animals.

13. Preadipocytes should be confluent to differentiate optimally into mature adipocytes. However, when combined with bioenergetic analyses using the XF Analyzer too many cells will lead to high basal respiration rates, making it problematic to evaluate the addition of compounds that induce respiration. In addition, too few preadipocytes will result in poor adipocyte differentiation and then function cannot be examined. Thus, it is important (through preliminary studies) to carefully determine the appropriate cell number (the minimal cell number preferentially) that can yield basal respiration rates that do not preclude the ability of activators of respiration (i.e., Norepinephrine, FCCP) to induce a significant increase in respiratory rate. We have established these cell numbers for inguinal and brown preadipocytes that can be used with high reproducibility following thawing of frozen stocks [4].
14. Supplementing XF media with 2 % BSA can scavenge free fatty acids and can limit lipolysis-mediated UCP1-independent mitochondrial uncoupling [9]. Make sure to pH the Seahorse DMEM after addition of BSA with NaOH.

References

1. Chance B, Williams GR (1956) The respiratory chain and oxidative phosphorylation. Adv Enzymol Relat Subj Biochem 17:65–134
2. Brand MD, Nicholls DG (2011) Assessing mitochondrial dysfunction in cells. Biochem J 435(2):297–312. doi:10.1042/BJ20110162
3. Gerencser AA, Neilson A, Choi SW, Edman U, Yadava N, Oh RJ, Ferrick DA, Nicholls DG, Brand MD (2009) Quantitative microplate-based respirometry with correction for oxygen diffusion. Anal Chem 81(16):6868–6878. doi:10.1021/ac900881z
4. Kazak L, Chouchani ET, Jedrychowski MP, Erickson BK, Shinoda K, Cohen P, Vetrivelan R, Lu GZ, Laznik-Bogoslavski D, Hasenfuss SC, Kajimura S, Gygi SP, Spiegelman BM (2015) A creatine-driven substrate cycle enhances energy expenditure and thermogenesis in beige fat. Cell 163(3):643–655. doi:10.1016/j.cell.2015.09.035
5. He J, Mao CC, Reyes A, Sembongi H, Di Re M, Granycome C, Clippingdale AB, Fearnley IM, Harbour M, Robinson AJ, Reichelt S, Spelbrink JN, Walker JE, Holt IJ (2007) The AAA+ protein ATAD3 has displacement loop binding properties and is involved in mitochondrial nucleoid organization. J Cell Biol 176(2):141–146. doi:10.1083/jcb.200609158
6. Yang MY, Bowmaker M, Reyes A, Vergani L, Angeli P, Gringeri E, Jacobs HT, Holt IJ (2002) Biased incorporation of ribonucleotides on the mitochondrial L-strand accounts for apparent strand-asymmetric DNA replication. Cell 111(4):495–505
7. Nicholls DG, Grav HJ, Lindberg O (1972) Mitochondrial from hamster brown-adipose tissue. Regulation of respiration in vitro by variations in volume of the matrix compartment. Eur J Biochem/FEBS 31(3):526–533
8. Parker N, Crichton PG, Vidal-Puig AJ, Brand MD (2009) Uncoupling protein-1 (UCP1) contributes to the basal proton conductance of brown adipose tissue mitochondria. J Bioenerg Biomembr 41(4):335–342. doi:10.1007/s10863-009-9232-8
9. Li Y, Fromme T, Schweizer S, Schottl T, Klingenspor M (2014) Taking control over intracellular fatty acid levels is essential for the analysis of thermogenic function in cultured primary brown and brite/beige adipocytes. EMBO Rep 15(10):1069–1076. doi:10.15252/embr.201438775

Chapter 13

Oxygen Consumption Rate and Energy Expenditure in Mice: Indirect Calorimetry

Eun Ran Kim and Qingchun Tong

Abstract

Global obesity epidemic demands more effective therapeutic treatments and better understanding of obesity pathophysiology. Since obesity results from energy imbalance, accurate quantification of energy intake and energy expenditure (EE) becomes an essential prerequisite to phenotype the cause for obesity development. Indirect calorimetry has long been used as one of the most established methods in EE quantification by detecting changes in levels of O_2 consumption and CO_2 production. In this article, we describe procedures and important considerations for an effective measurement using indirect calorimetry.

Key words Indirect calorimetry, Energy expenditure, Metabolic rate, Oxygen consumption, Obesity

1 Introduction

During past several decades, obesity has been increasing at an alarming rate, and it has been recognized that obesity is one major contributing factor to multiple diseases, including coronary heart disease, hyperlipidemia, hypertension, stroke, and some forms of cancers [1–3]. Thus, research on the cause for the obesity development is imperative. Obesity is a condition of imbalance between energy intake and energy expenditure (EE, thermogenesis) [4]. An assessment of energy intake is relatively simple and can be done by measuring amount of food being consumed (or calorie consumption). However, it appears to be more complicated to measure EE, which is estimated by the amount of O_2 consumption and CO_2 production during oxidation of nutrients [5].

In 1780, Lavoisier used for the first time the term "calorimeter," which quantifies heat production after inhalation of oxygen as calorie [6]. To measure heat production as EE, he developed the first calorimeter in animals. Since then, through many trials of building a prototype calorimeter, a more simple, researcher-friendly, and practical system has been developed. The early model relied upon direct calorimetry to measure EE. Although this

Jun Wu (ed.), *Thermogenic Fat: Methods and Protocols*, Methods in Molecular Biology, vol. 1566,
DOI 10.1007/978-1-4939-6820-6_13,

method has a high level of accuracy in EE measurement, which is directly calculated from heat production of a body, it failed to reach popularity among researchers. The reasons behind the method's low preference for measuring EE include its high cost and considerable time required for comprehensively learning and understanding the system before utilizing it [7]. Another model is the indirect calorimetry (IC) measurement. IC does not require expert technical knowledge, is inexpensive, and is able to perform simultaneous measurements on multiple numbers of animals in one experiment. There are two types of systems for IC: closed-circuit system and open-circuit system [8]. The open-circuit system has been widely used among researchers primarily in small animals such as rats or mice [9–11]. In this system, continuous gas flow of dried room air passes through a sealed chamber. O_2 and CO_2 sensors measure their respective concentrations from incurrent and excurrent dried air. The difference in O_2 and CO_2 levels of the exchanged gas from the chamber is considered O_2 consumption and CO_2 production, respectively [7]. Using IC in various conditions makes it feasible for researchers to estimate metabolic rate, for example, in normal chow vs. high diet-fed conditions, during different nutritional status (fed vs. fasted), in response to drugs, and in different genetic mouse models [12–14]. Here, we show how IC can be performed using an open-circuit indirect calorimetry method and discuss important considerations utilizing this method in the following Subheading 4.

2 Materials

2.1 Animals

1. Mice: male C57BL/6J mice (10–12 week-old, 25–26 g body weight), male FVB/N mice (15–16 week-old, average 38 g body weight), β_1, β_2, β_3-adrenergic receptors global knock-out (*Beta-less*) mice (15–16 week-old, average 37 g body weight) were used for EE measurements [15].
2. Housing conditions for mice: room temperature was maintained at 22 ± 1 °C and mice were kept in 12:12 h light and dark cycles (*see* **Note 1**).
3. Diet: Mice were fed standard chow diet (Teklad F6 Rodent Diet 8664, 4.05 kcal/g, 3.3 kcal/g metabolizable energy, 12.5 % kcal from fat, Harlan Teklad, Madison, WI) or high sucrose/high fat diet (Research diets, D12331, 25.5 kcal% of carbohydrate, 58.0 kcal% of fat).

2.2 Open-Circuit Indirect Calorimetry System

1. Indirect calorimetry apparatus (Columbus Instruments Comprehensive Lab Animal Monitoring System, CLAMS) with a computer containing Oxymax software was used for operating the system. The system is equipped with chambers

for housing mice, feeders, water bottles, scales for monitoring food intake, oxygen and carbon dioxide sensors, *X* and *Y*-Axis activity sensors, and a sample dryer (*see* **Note 2**).

2. Desiccant (Drierite) was used to dry air. Gases (O_2; 20.50 % and CO_2; 5000 ppm balanced with nitrogen, Praxair) were used for calibration.

3 Methods

3.1 Preparation of the Calorimetry Apparatus with Animals and Running the Experiment

1. Place water bottles, feeders, individual housing chambers, and desiccant onto the calorimetry system.
2. Power on the CLAMS and computer connected to it.
3. Start the program and initialize the calorimetry system.
4. Wait for minimum 2 h for stabilizing the system before calibration.
5. Start calibration for O_2 and CO_2 sensors: Adjust sample flow 0.4–0.5 mL/min and connect to a calibration gas tank. Open the gas before calibrating the sensors. Actual gas composition may vary from requested concentration. Use actual O_2 concentration as the reference gas.
6. Set O_2 % of the sensor to the reference gas % using a gain knob of the O_2 sensor (*see* **Note 3**).
7. Start calibration for the CO_2 sensor. Adjust offset gas using a knob on CO_2 sensor until the numbers match the actual gas concentration. Set up the experiment. In this step, animal ID and body weight are registered with an assigned chamber. Also, measurement timing of cage settle, cage measure, reference settle, reference setting is defined at this step (*see* **Note 4**).
8. Save it as a data file (*see* **Note 5**).
9. Place mice into chambers, one in each individual chamber (*see* **Note 6**).
10. Close covers tightly (*see* **Note 7**).
11. Run the experiment (*see* **Note 8**).
12. Finish the experiment.
13. Return mice to their home cages after checking body weight and health condition (*see* **Note 9**).
14. Export data collected as excel files for further analysis (*see* **Note 10**).

3.2 Data Interpretation

O_2 consumption, CO_2 production, respiratory exchange ratio (RER), heat production, locomotion activity, and feeding data can be obtained as follows:

1. *Oxygen consumption (VO2) = rates of VO2 flow at the input (ViO2i)–output (VoO2o), CO2 production (VCO2) = rates of CO2 flow at the output (VoCO2o)–input (ViCO2i), respiratory exchange ratio (RER) = VCO2/VO2, heat production = (3.815+1.232 × RER) × VO2* (*see* **Note 11**).
2. Activity level is counted as infrared beam breaks on an X-Y axis. It can be broken down into two types of activity levels: (1) locomotion activity in which the program counts horizontal distinct *X-Y* beam breaks and (2) spontaneous activity in which the program counts the same beam breaks. Those separate data are collected automatically.
3. Feeding is calculated as food consumption at the time metabolic data are logged (*see* **Note 12**).
4. Mice are nocturnal animals, and their feeding and activities peak during night periods (Fig. 1a, b). O_2 consumption, CO_2 production, and heat production are dependent on food intake and locomotion activity (Fig. 1c–e). RER values increase during nighttime indicating that a switch from lipid to carbohydrate as a nutritional source (Fig. 1f).
5. Energy expenditure is increased by switching from regular chow to high fat diet (Fig. 2a). RER value clearly shows a transition to high fat diet with a reduction close to 0.7 (Fig. 2b) (*see* **Note 13**).
6. Lack of β-adrenergic receptors reduced diet-induced energy expenditure, especially during night periods (Fig. 3). Feeding induces sympathetic activation followed by increased energy expenditure and thermogenesis. A major part of the effect is mediated through β-adrenergic receptors [15]. The effect on energy expenditure is diminished in β-less mice and its deficiency is well detected using an indirect metabolic system (*see* **Note 14**).

4 Notes

1. Animal care and procedures need to be performed following guidelines and regulations approved by research institutional animal care and use committee.
2. The environment for CLAMS should be similar to regular housing conditions including humidity and temperature to minimize stress from changing environment. It is best that the room housing the system is under tight control of environmental changes including noise and number of people to enter at any given time. Any disturbance can contribute to metabolic changes of animals, resulting in alterations in the levels of O_2 consumption (EE).

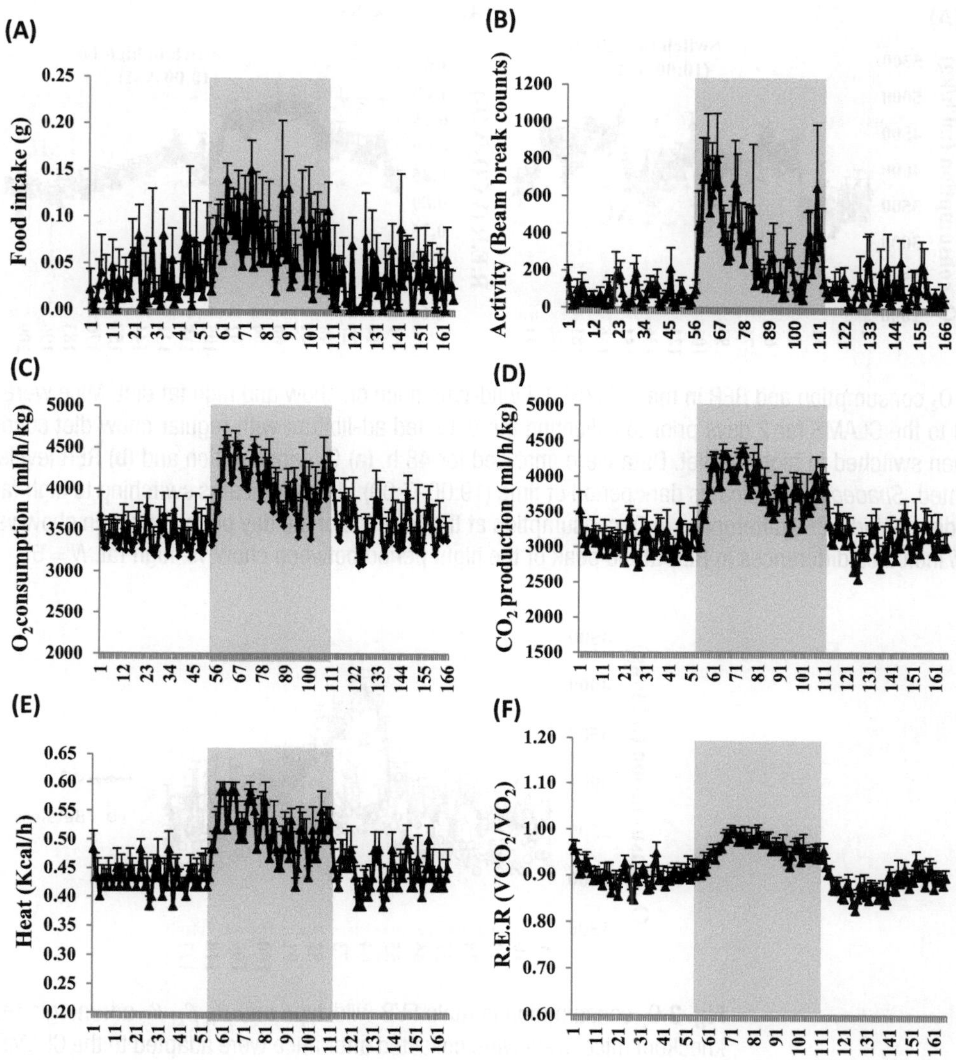

Fig. 1 Metabolic rates in male C57BL/6J wild-type mice. Mice were acclimated to the CLAMS for 2 days prior to collecting the data; animals were fed ad-libitum with regular chow diet. Data were analyzed for 36 h. (**a**) food intake, (**b**) locomotion activity levels, (**c**) O_2 consumption, (**d**) CO_2 production, (**e**) heat production, and (**f**) RER were presented. Shaded area indicates dark period of time (19:00–7:00). RER: respiratory exchange ratio. $N = 5$

3. The actual gas % from the calibration tank is referred to as reference gas.
4. This setting allows a scheduled amount of time for the system to measure gas sample from each chamber. The system samples gas from one chamber before moving onto another. During transition, gas from the previous chamber purged to the sensors will be washed out. If a given time is too short for each segment to purge gas, the gas sample may become contaminated from the previous chamber sample, or the system may not be able to collect stable actual readings. If a given time is

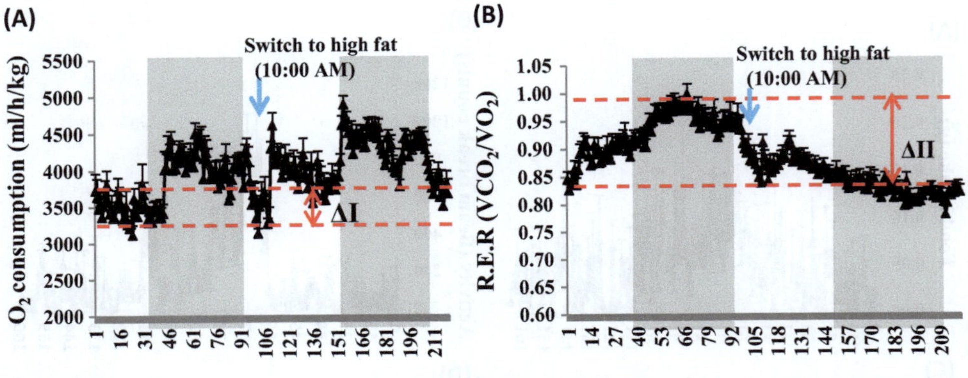

Fig. 2 O_2 consumption and RER in male C57BL/6J wild-type mice on chow and high fat diet. Mice were acclimated to the CLAMS for 2 days prior to collecting the data, fed ad-libitum with regular chow diet conditions, and then switched to high-fat diet. Data were analyzed for 48 h. (**a**) O_2 consumption and (**b**) RER levels were presented. *Shaded area* indicates dark period of time (19:00–7:00). *Arrow* indicates switching to high fat from chow diet. *ΔI* indicates differences in O_2 consumption at the baseline of the day period between chow vs. high fat. *ΔII* indicates differences in RER at the peak of the night period between chow vs. high fat. $N = 5$

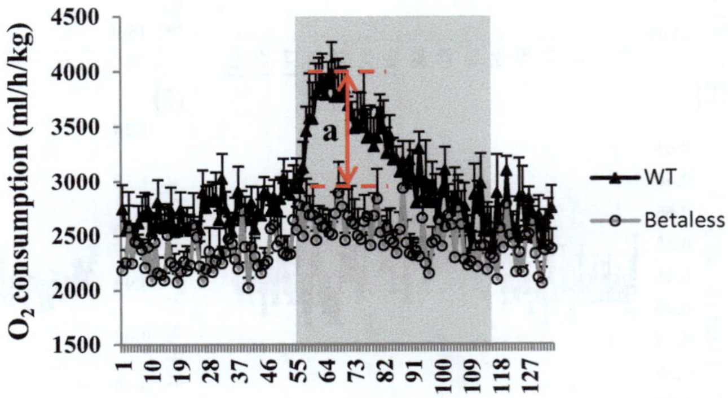

Fig. 3 O_2 consumption in male FVB wild-type and β_1, β_2, β_3-adrenergic receptor knockout mice. Data were collected after mice were adapted to the CLAMS environment. Mice had free access to regular chow diet. "*a*" indicates different levels of O_2 consumption between two different mouse genotypes during night periods. *Shaded area* indicates dark period of time (19:00–7:00). $N = 4$–5/group

too long for each segment, it will reduce frequencies of sampling from each chamber. This can produce a crude measurement. It is also required to set a frequency of reference gas measurements at a certain frequency, which depends on room environment. Unless the environment is very controlled and stable for the CLAMS, it is recommended to measure reference more than once per round measurement of all chambers.

5. Data is automatically saved while the experiment is running in real time.
6. A centered feeder model, the most commonly used one, has a Barbie table, to prohibit mouse from sitting on a feeder. This

allows running of the experiment with both normal and obese mice. There are different sizes of spacers which reserve space between the feeder and the table. Depending on the size of the animal, an optimally sized spacer needs to be used. The size of the spacer is chosen to disallow the mouse to rest on the feeder but allow enough space for food access.

7. CLAMS measures EE based on gas in and out from the chamber. The delta between in and out of gas derives O_2 consumption and CO_2 production levels. Since mouse size is a small, the gas exchange has to be precisely captured. Thus, it is critical to maintain the chamber without leakage while running experiment.
8. (a) CLAMS chamber size is smaller than a regular housing cage, and it is enclosed, which might cause stress responses to animals. Thus, an acclimation period is recommended to eliminate the concerns of changes in O_2 consumption caused by stress. The length of running an experiment should include the acclimation period. (b) Monitor mice condition during acclimation periods by monitoring food intake and behaviors daily. (c) Generally, mice have an acclimation time arranging from a couple of hours to a few days. For data analysis, the data collected after the acclimation period may be used.
9. It is important to check the health condition of animals since it may directly and dramatically influence metabolic rate. If animals lose more than 10 % of body weight during experiment, the subject should be considered for removal from the group. Mice having a daily body weight change less than 10 % are normal.
10. Data collected during the period of known interruption on the chambers will not be used in analysis, such as when the chamber lid was opened to give injections, or to switch different types of food. These procedures wash out the air in chambers.
11. RER normally varies between 0.7 and 1. It is "1" if the animal uses only carbohydrate as nutritional sources and "0.7" only fat. Since mice consume a mixture of nutrients, RER value varies between 1 and 0.7 [16, 17].
12. Meal frequency and duration can also be extracted.
13. CLAMS data automatically normalizes to body weight of animal. However, this could be misleading in some situations, i.e., comparing lean vs. obese mice. Brown adipose is a major tissue of EE, but white adipose is not metabolically active like liver or skeletal muscle. Therefore, if bigger body size is due to white fat deposition, it would be misleading if VO_2 or VCO_2 is normalized to body weight. However, it is also sometimes misleading if VO_2 or VCO_2 is normalized to lean mass, as white fat mass may contribute significantly to EE, which has been

increasingly supported by the fact that some white fat cells are capable of turning into brown fat cells [18]. One way to overcome this pitfall is to conduct O_2 measurement prior to obesity development, which will eliminate the concerns caused by body weight difference. Alternatively, as recently proposed by others [19, 20], the collected data can be analyzed with regression analysis, which appears to be able to obtain consistent comparison between study groups regardless of body weight, fat, or lean mass.

14. CLAMS measurement can be used to study the impact of different drugs on EE. CLAMS measurement can be a powerful paradigm to examine the acute effect on EE by the drug of interest. We have successfully used this system to examine the effect of the melanocortin receptor 4 agonist MTII [9] and the effect of high-fat diet (Fig. 2) to increase EE. Since EE normally shows a diurnal pattern with higher levels during the dark cycle and depends on feeding status with lower levels during fasting, timing for drug administration should be considered. This consideration may be based on the expected effect of the drug of interest. For a drug expected to increase EE, it can be administered during fasting, and for that to reduce EE, it can be administered during the dark cycle.

Acknowledgment

This work was supported by NIH R01DK092605. Q.T. is the holder of Cullen Chair in Molecular Medicine and Welch Research Scholar (L-AU0002) of the University of Texas McGovern Medical School.

References

1. Ryan KK, Woods SC, Seeley RJ (2012) Central nervous system mechanisms linking the consumption of palatable high-fat diets to the defense of greater adiposity. Cell Metab 15(2):137–149
2. Heitmann BL, Westerterp KR, Loos RJ, Sorensen TI, O'Dea K, McLean P, Jensen TK, Eisenmann J, Speakman JR, Simpson SJ, Reed DR, Westerterp-Plantenga MS (2012) Obesity: lessons from evolution and the environment. Obes Rev 13(10):910–922
3. Despres JP, Lemieux I (2006) Abdominal obesity and metabolic syndrome. Nature 444(7121):881–887
4. Jung RT, Shetty PS, James WP, Barrand MA, Callingham BA (1979) Reduced thermogenesis in obesity. Nature 279(5711):322–323
5. Frankenfield DC (2010) On heat, respiration, and calorimetry. Nutrition 26(10):939–950
6. Roberts L (1991) A word and the world: the significance of naming the calorimeter. Isis 82:199–222
7. Levine JA (2005) Measurement of energy expenditure. Public Health Nutr 8(7A):1123–1132
8. Maclagan NF, Sheahan MM (1950) The measurement of oxygen consumption in small animals by a closed circuit method. J Endocrinol 6(4):456–462
9. Xu Y, Wu Z, Sun H, Zhu Y, Kim ER, Lowell BB, Arenkiel BR, Xu Y, Tong Q (2013) Glutamate mediates the function of melanocortin receptor 4 on Sim1 neurons in body weight regulation. Cell Metab 18(6):860–870

10. Kong D, Tong Q, Ye C, Koda S, Fuller PM, Krashes MJ, Vong L, Ray RS, Olson DP, Lowell BB (2012) GABAergic RIP-Cre neurons in the arcuate nucleus selectively regulate energy expenditure. Cell 151(3):645–657
11. Rezai-Zadeh K, Yu S, Jiang Y, Laque A, Schwartzenburg C, Morrison CD, Derbenev AV, Zsombok A, Munzberg H (2014) Leptin receptor neurons in the dorsomedial hypothalamus are key regulators of energy expenditure and body weight, but not food intake. Mol Metab 3(7):681–693
12. Rothwell NJ, Stock MJ (1982) Energy expenditure of 'cafeteria'-fed rats determined from measurements of energy balance and indirect calorimetry. J Physiol 328:371–377
13. Enerback S, Jacobsson A, Simpson EM, Guerra C, Yamashita H, Harper ME, Kozak LP (1997) Mice lacking mitochondrial uncoupling protein are cold-sensitive but not obese. Nature 387(6628):90–94
14. Crane JD, Palanivel R, Mottillo EP, Bujak AL, Wang H, Ford RJ, Collins A, Blumer RM, Fullerton MD, Yabut JM, Kim JJ, Ghia JE, Hamza SM, Morrison KM, Schertzer JD, Dyck JR, Khan WI, Steinberg GR (2015) Inhibiting peripheral serotonin synthesis reduces obesity and metabolic dysfunction by promoting brown adipose tissue thermogenesis. Nat Med 21(2):166–172
15. Bachman ES, Dhillon H, Zhang CY, Cinti S, Bianco AC, Kobilka BK, Lowell BB (2002) betaAR signaling required for diet-induced thermogenesis and obesity resistance. Science 297(5582):843–845
16. Dewar AD, Newton WH (1948) The relationship between food intake and respiratory quotient in mice. Br J Nutr 2(2):142–145
17. Guo J, Hall KD (2009) Estimating the continuous-time dynamics of energy and fat metabolism in mice. PLoS Comput Biol 5(9):e1000511
18. Cohen P, Spiegelman BM (2015) Brown and beige fat: molecular parts of a thermogenic machine. Diabetes 64(7):2346–2351
19. Tschop MH, Speakman JR, Arch JR, Auwerx J, Bruning JC, Chan L, Eckel RH, Farese RV Jr, Galgani JE, Hambly C, Herman MA, Horvath TL, Kahn BB, Kozma SC, Maratos-Flier E, Muller TD, Munzberg H, Pfluger PT, Plum L, Reitman ML, Rahmouni K, Shulman GI, Thomas G, Kahn CR, Ravussin E (2012) A guide to analysis of mouse energy metabolism. Nat Methods 9(1):57–63
20. Kaiyala KJ, Schwartz MW (2011) Toward a more complete (and less controversial) understanding of energy expenditure and its role in obesity pathogenesis. Diabetes 60(1):17–23

Chapter 14

Isolation and Patch-Clamp of Primary Adipocytes

Yanhui Zhang, Dan Tong, Anil Mishra, Litao Xie, Isaac Samuel, Jessica K. Smith, and Rajan Sah

Abstract

The patch-clamp technique allows for the study of ion channel activity in the native adipocyte environment to better understand the contributions of ion channels to adipocyte signaling. Here, we describe methods for isolating primary mature adipocytes from both mouse and human white adipose tissues (subcutaneous and visceral). From the same preparation, we describe how to culture and differentiate preadipocytes isolated from the stromal vascular fraction. We then describe in detail patch-clamp methods, including both whole-cell and perforated-patch configurations.

Key words Patch-clamp, Adipocyte isolation, Culture, Differentiation, Whole-cell patch-clamp, Perforated-patch

1 Introduction

Obesity is a growing public health problem that predisposes individuals to health conditions such as diabetes and heart disease. The obesity epidemic has inflicted healthcare costs over 100 billion dollars in the United States alone. For this reason, there is a growing need for investigators to become aware of novel methods to study adipocytes. Like other cells types, adipocytes express ion channels on the plasma membrane [1]. These ion channels are important for adipogenesis [1], fatty acid sensing [2], oxidative metabolism, inflammation, and energy homeostasis [3]. For this reason, developing sensitive assays to measure ion channel activity in primary adipocytes will provide a deeper understanding of adipocyte signaling. The patch-clamp technique has been used for decades to study ion channel signaling in numerous different cell types [4–7], providing an approach to examine signaling within the native cellular environment. We show how the patch-clamp technique can be applied to adipocytes isolated from mouse or human adipose tissue; both freshly isolated mature adipocytes and cultured primary adipocytes.

Jun Wu (ed.), *Thermogenic Fat: Methods and Protocols*, Methods in Molecular Biology, vol. 1566, DOI 10.1007/978-1-4939-6820-6_14, © Springer Science+Business Media LLC 2017

2 Materials

2.1 Murine and Human Primary Adipocyte Isolation

1. Digestion buffer: HBSS (1×), 3 % BSA, 1 mM $CaCl_2$, 1 mM $MgCl_2$, 8 mg/mL collagenase D, 2.4 units/mL Dispase II.
2. Media: 90 % DMEM/F12 medium, 10 % FBS, 100 IU penicillin, and 100 μg/mL streptomycin.
3. Animal surgery supplies: forceps, small scissors and razor blades.
4. 50 mL conical centrifuge tubes, benchtop centrifuges.
5. Tissue culture supplies: petri dishes and plates, collagen coated cell culture plates, glass coverslips, incubator.
6. Nylon mesh filters with pore size of 220 μm.
7. Corning Matrigel matrix.

2.2 Murine and Human Primary Adipocyte Culture and Differentiation

1. Murine adipocyte culture medium: 90 % DMEM/F12 medium, 10 % FBS, 100 IU penicillin, and 100 μg/mL streptomycin.
2. Murine adipocyte differentiation cocktail: 0.5 mM IBMX, 1 μM Dexamethasone, 850 nM Insulin, 1 μM Rosiglitazone.
3. Human adipocyte culture medium: 90 % Preadipocyte Basal Medium 2 (PBM-2) (Lonza, PT-8202), 10 % FBS, l-glutamine, gentamycin, and amphotericin (Lonza, PT9502).
4. Human adipocyte differentiation cocktail: Dexamethasone, IBMX, indomethacin, and human insulin (Lonza, PT-9502).
5. Sterile PBS.
6. Tissue culture plates, petri dishes, and incubator.

2.3 Patch Clamp

1. Electrophysiology required equipment: Anti-vibration table with a Faraday cage, Microscope, Axon Digidata 1550 Low-Noise Data Acquisition System, Axopatch 200 B amplifier or a MultiClamp 700 B amplifier, Headstage, Micromanipulator, Borosilicate glass capillary holder, Pressure control system, Computer, Recording software, Camera, Glass pipette puller, Micro forge, Osmometer.
2. Consumables: Borosilicate glass capillaries, Three-way valve, Tubes for perfusion and pressure system, 0.2 μm filter.
3. Solutions: All solutions are prepared using ultrapure water (18.2 MΩ-cm at 25 °C). Choose the bath and pipette solutions that are optimal for isolating your current of interest.

3 Methods

3.1 Murine and Human Primary Adipocyte Isolation

1. *Mouse:* Remove mouse inguinal or epididymal white adipose (iWAT or eWAT) pads.
   Human: Harvest fresh human visceral adipose tissues samples (~1 cm^3) from esophageal fat pads (e.g., during intra-abdominal

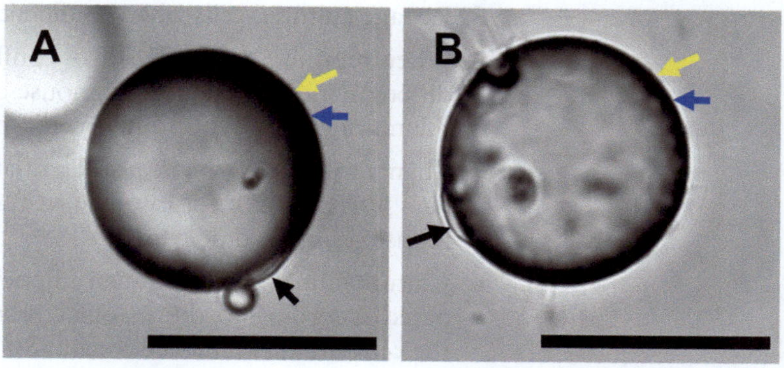

Fig. 1 (**a**) Isolated mouse primary mature adipocyte. (**b**) Isolated human visceral primary mature adipocyte. *Black arrow*: nucleus, *yellow arrow*: membrane, *blue arrow*: lipid droplet (Scale bar: 50 μm)

laparoscopic surgery) of bariatric (BMI >30 kg/m^2) or non-bariatric surgery patient (BMI <30 kg/m^2), place immediately on ice, and transport to laboratory within 20 min (*see* **Note 1**). Handling of the adipose sample during surgery should be gentle, restricting use of cautery or other energy sources to the peritoneum only and not the adipose cells.

2. Wash adipose tissue with HBSS, and then mince with a razor blade.
3. Digest the minced tissue for 30–40 min at 37 °C in 10 mL digestion buffer.
4. Filter the digested tissue through a 220 μm mesh, and centrifuge at 300 × *g* for 3 min.
5. Collect supernatant (this fraction contains mature adipocytes).
6. Wash cells with media by repetitive pelleting (*see* **Note 2**).
7. For mature adipocytes, aliquot the final adipocyte suspension into a 24-well culture dish.
8. Coat glass coverslips with Matrigel (*see* **Note 3**) and place them upside down on the surface of the cell suspension. Incubate at 37 °C for at least 20 min (*see* Fig. 1 for an example of freshly isolated mature adipocyte).
9. To isolate the stromal vascular fraction (SVF), save the pellet from **step 4** and wash it with media by repetitive pelleting.
10. Seed SVF on collagen coated plates for subsequent culture and differentiation.

3.2 Murine and Human Primary SVF Culture and Differentiation

1. Seed mouse or human SVF on collagen coated plates and culture in the murine adipocyte culture medium listed above.
2. After 24 h, wash cells with PBS and shake vigorously to remove debris.

3. Passage cultured cells when confluence reaches 80 %.
4. When murine SVF reach 100 % confluence, induce adipocyte differentiation by adding the mouse differentiation cocktail to the murine adipocyte culture medium.
5. Maintain the cells in the differentiation medium for 10–12 days (*see* Fig 2a, b for example of cultured and differentiated mouse primary adipocytes).
6. Culture human SVF in human adipocyte culture medium (prepared according to the manufacturer's instructions). Passage the cells when confluence reaches 80 %.
7. To induce human primary adipocyte differentiation, prepare differentiation medium by adding the human differentiation cocktail to human adipocyte culture medium according to the manufacturer's instructions (Lonza, MD).
8. Maintain human primary adipocytes in differentiation medium for 10–12 days, or until droplets formation (*see* Fig. 2c).

3.3 Mature Adipocyte: Whole-cell Patch-Clamp

All procedures are conducted at room temperature unless specified.

1. Place the coverslip with mature adipocytes in the recording chamber (*see* **Note 4**).
2. Use a glass capillary puller to make pipettes from borosilicate glass with a resistance of 3–4 MΩ when filled with internal solution.
3. Advance the glass pipette filled with pipette solution into the bath and toward the adipocyte. Apply a very light positive pressure to the pipette.

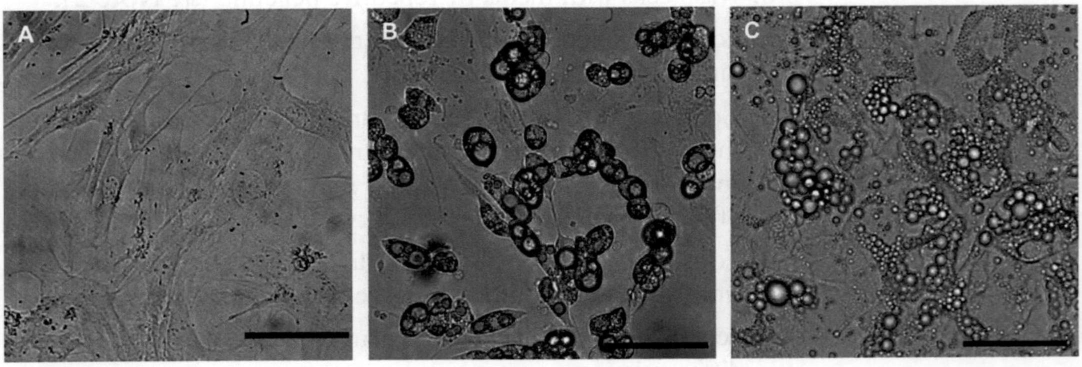

Fig. 2 (**a**) Cultured primary murine pre-adipocytes isolated from SVF. (**b**) Differentiated primary murine adipocytes. (**c**) Differentiated human visceral primary adipocytes (Scale bar: 200 μm)

4. Approach the patch-pipette to the adipocyte to patch and correct the pipette offset.
5. Advance the glass pipette until the tip touches the adipocyte.
6. Release the positive pressure and apply gentle negative pressure to obtain a GΩ seal.
7. Once the GΩ seal has been formed, apply light and short suction pulses by a syringe or by mouth to rupture the membrane below the patch. Lipid will enter into the pipette tip immediately after patch-rupture. Apply light and short positive pressure to push the lipid out of the pipette tip taking care not to break the GΩ seal.
8. Perform cell capacitance and series resistance compensation.
9. Recordings can be acquired and analyzed using appropriate software.

3.4 Mature Adipocyte: Perforated Patch-Clamp

1. Prepare a stock solution of amphotericin B at a concentration of 60 mg/mL by dissolving in DMSO. Keep stock solution away from light. Store the stock solution at room temperature.
2. Add the stock solution to 1 mL of internal pipette solution and mix to achieve a final concentration of 360 μg/mL. (Optional Step: Add fresh amphotericin to the pipette solution every 4–5 h).
3. Fill a borosilicate pipette halfway with the internal pipette solution from **step 2**.
4. Tap the pipette a few times to eliminate any air bubbles that might be present in the tip of the pipette.
5. Secure the pipette in the pipette holder and place the pipette tip into the external solution in the recording chamber. Move the pipette tip toward the adipocyte as rapidly as possible. Do not apply positive pressure.
6. Gently push the pipette tip onto the mature adipocyte membrane.
7. Apply gentle negative pressure/suction to obtain a GΩ seal (*see* Fig. 3).
8. The electrical access takes several minutes to obtain and can be monitored by tracking the capacity transient.
9. As soon as the capacity transient stabilizes and access resistance (R_a) is less than 15 MΩ, cell capacitance and series resistance compensation can be carefully performed.
10. Access resistance (R_a) should be monitored during the experiment. (Example: In our experiments, data was only collected when this value was stable and less than 15 MΩ).

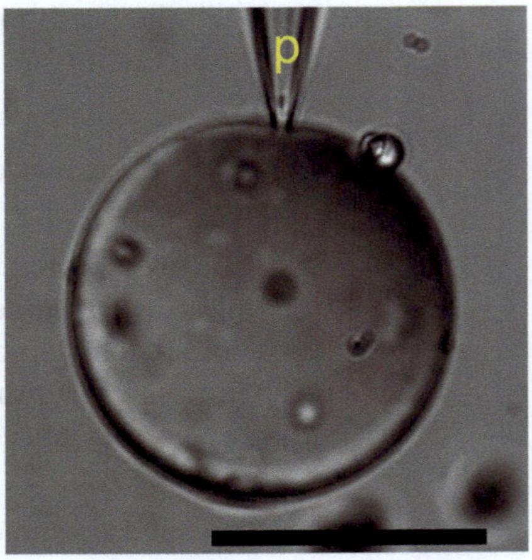

Fig. 3 Perforated patch-clamp of a mature freshly isolated adipocyte from mouse iWAT. *P* patch pipette. (Scale bar: 50 μm)

4 Notes

1. Human adipose tissue should be obtained under your institution's regulation, ensuring prior Institutional Review Board approval and surgical patient's Informed Consent.
2. Repetitive Pelleting Procedure: add more media to cells suspension, gently pipette cells to wash them in media, centrifuge again, and collect supernatant. Repeat this for 2–3 times.
3. Warm up Matrigel to 37 °C before the experiment, then quickly apply a small amount of the Matrigel to a glass coverslip and spread with a pipette tip. Let it air dry at room temperature.
4. Flip the coverslip, so that the side with adipocytes will face upward in the recording chamber.

Acknowledgment

This work was supported by a grant from the American Heart Association and from the Roy Carver Charitable Trust to R.S.

References

1. Che H, Yue J, Tse HF, Li GR (2014) Functional TRPV and TRPM channels in human preadipocytes. Pflugers Arch 466(5):947–959
2. Sukumar P et al (2012) Constitutively active TRPC channels of adipocytes confer a mechanism for sensing dietary fatty acids and regulating adiponectin. Circ Res 111(2):191–200
3. Ye L et al (2012) TRPV4 is a regulator of adipose oxidative metabolism, inflammation, and energy homeostasis. Cell 151(1):96–110
4. Gardner P (1990) Patch clamp studies of lymphocyte activation. Annu Rev Immunol 8:231–250
5. Molleman A (2003) Patch clamping: an introductory guide to patch clamp electrophysiology. Wiley, New York
6. Sakmann B, Neher E (1984) Patch clamp techniques for studying ionic channels in excitable membranes. Annu Rev Physiol 46:455–472
7. Hille B (2001) Ion channels of excitable membranes, 3rd edn. Sinauer Associates, Sunderland, MA

Chapter 15

In Vitro Approaches to Model and Study Communication Between Adipose Tissue and the Liver

Sean O'Connor and Paul Cohen

Abstract

Inter-organ communication is central to mammalian metabolism, allowing for the coordination of crucial homeostatic processes such as food intake, body weight, and blood glucose. Dissecting this process at a cellular and molecular level requires the establishment of a reductionist system in which cell types of interest can be investigated as the source and target of biologically relevant signals. Here, we describe a system to study the interaction between adipocytes and hepatocytes using either conditioned media from cultured adipocytes or co-culture of hepatocytes with adipose explants. These methods have the potential to identify novel polypeptides and metabolites involved in the adipose-liver metabolic axis.

Key words Hepatocytes, Adipocytes, Conditioned media, Tissue explants, Signaling factors, Co-culture, Adipokines

1 Introduction

Adipocytes or fat cells are highly efficient at storing energy in the form of triglycerides and can rapidly mobilize these stores at times when nutrients are scarce. Adipose tissue is also an important endocrine organ, secreting polypeptides (adipokines) or metabolites that can act on distant organs to regulate systemic metabolism [1]. Well-characterized adipokines include leptin, adipsin, adiponectin, resistin, and RBP4 among others [2–6]. Genomic and proteomic studies suggest that adipocytes can secrete at least several hundred distinct proteins [7, 8], and for the vast majority their target tissue and biological action remains unknown.

Genetically engineered mouse models frequently point to communication between adipocytes and other organs such as liver, muscle, or pancreatic islets. For example, deletion of the transcriptional coregulator PRDM16 in adipocytes results in hepatic insulin resistance and hepatic steatosis, but the relevant signal from fat to liver responsible for this phenotype is unclear [9]. One approach to studying and identifying the putative signals that mediate

Jun Wu (ed.), *Thermogenic Fat: Methods and Protocols*, Methods in Molecular Biology, vol. 1566,
DOI 10.1007/978-1-4939-6820-6_15, © Springer Science+Business Media LLC 2017

inter-organ communication involves establishing a cell culture system to study the effects of conditioned medium from one cell type, such as an adipocyte, on a biological readout in a target cell type, such as a hepatocyte.

Here, we describe methods using cultures of primary adipocytes and hepatocytes or adipose tissue explants to study adipose-liver communication. We detail methods used to isolate and culture primary hepatocytes with either adipocyte conditioned media or fat tissue explants in a co-culture system. In the notes, we also review technical concerns that we have encountered and optimized. This type of approach could in principle be applied to the study of any cell-cell communication model.

2 Materials

Prepare all media, buffers, and solutions in a laminar flow hood. Surgeries for explant and hepatocyte isolation can be carried out on a benchtop; however, be sure to spray all tools and surfaces with 70 % ethanol to reduce risk of contamination.

2.1 Collection of Primary Adipocyte Conditioned Medium

1. Collection Medium: DMEM (4.5 g/L glucose, including glutamine) with 0.2 % BSA, 2 mM sodium pyruvate, 1 % penicillin/streptomycin, 0.1 μM dexamethasone, and optional 1 nM insulin (*see* **Note 1**). Pass through a 0.2 μm filter before use.

2.2 Preparation of Fat Tissue Explants (See Note 2)

1. Mice of the genotype, sex, and age of choice (*see* **Note 3**).
2. Milligram scale (*see* **Note 4**).
3. A dissection board and dissection tools including scissors and forceps, pins, 70 % ethanol, cold 1× sterile PBS, and several sterile dishes to place freshly isolated tissue.
4. Explant culture medium: DMEM/F12 + GlutaMAX (Life Technologies), 10 % fetal bovine serum, 1 % penicillin/streptomycin.
5. Co-culture inserts designed for 12-well plates (*see* **Note 5**).

2.3 Primary Hepatocyte Isolation

1. 8–12 week old C57BL/6 mice (*see* **Note 6**).
2. Seeding medium: DMEM (4.5 g/L glucose, including glutamine) with 10 % FBS, 2 mM sodium pyruvate, 1 % penicillin/streptomycin, 1 μM Dexamethasone, and 100 nM insulin.
3. Culture medium: Collection medium (*see* Subheading 2.1) with 1 nM insulin added (*see* **Note 1**).
4. Perfusion buffer: Hank's Balanced Salt Solution with 0.4 g/L KCl, 1 g/L glucose, 2.1 g/L sodium bicarbonate, and 0.2 g/L EDTA. Pass through a 0.2 μm filter before use.

5. Liver digest buffer: Collagenase-Dispase solution (*see* **Note 7**), adjusted to pH 7.4 and passed through a 0.2 μm filter before use (*see* **Note 8**).
6. Percoll solution: 90 % Percoll with 10 % 10X PBS. Pass through a 0.2 μm filter before use.
7. Peristaltic pump adjusted to a flow rate of 2.5 mL/min with a small bubble catcher fitted to tubing (*see* **Note 9**).
8. Butterfly needle (23G × 3/4″) to be attached to the output end of tubing.
9. A dissection board and dissection tools including scissors, forceps, and pins.
10. 42 °C water bath located adjacent to the surgical area.
11. 50 mL Falcon tubes with fitted 100 μm filters, 1X sterile PBS, and sterile dishes for placing freshly isolated tissue.
12. Syringe with prepared anesthetic (*see* **Note 10**).
13. Collagen coated tissue culture plates (*see* **Note 11**).

3 Methods

3.1 Collection of Primary Adipocyte Conditioned Medium

1. Following full differentiation of primary adipocytes, gently wash cells twice with PBS, then add collection medium to each well (*see* **Note 12**).
2. After 24 h, transfer conditioned medium to microcentrifuge tubes and centrifuge for 10 min at 4000 × *g*.
3. Remove supernatant and either use immediately to treat hepatocytes (*see* Subheading 3.4), or store at −80 °C in aliquots.

3.2 Preparation of Fat Tissue Explants

1. After securing a freshly euthanized mouse on the dissection board, pull the skin of the abdomen away from the body and make an incision through the skin without piercing the peritoneum.
2. Cut the skin from the hips up toward the shoulders, then along each arm and leg. Peel back the skin to reveal the inguinal fat between the peritoneum and skin, then pin the skin to the dissection board.
3. Remove the large lymph node located in each inguinal fat depot, then excise both depots, and place in sterile PBS on ice (Fig. 1).
4. Working in a tissue culture hood, use surgical scissors to cut each depot into small pieces, about 5–10 mg each (*see* **Note 13**).
5. Place a small dish with sterile PBS on the scale, then transfer 100 mg of explants (10–20 pieces) into the dish.

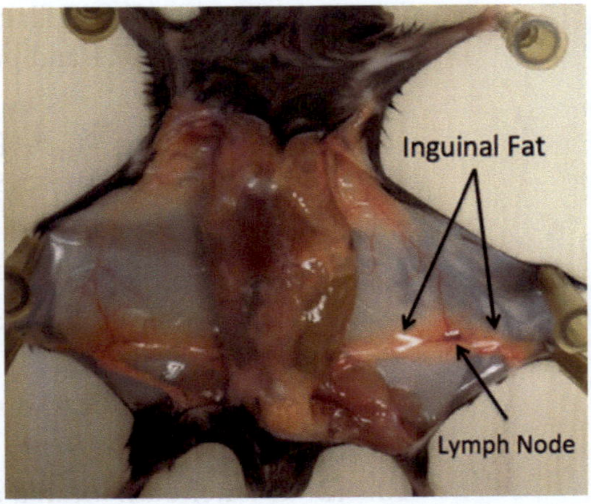

Fig. 1 Surgical access to the inguinal depot. The peritoneum should be left intact as the skin is peeled away to reveal the subcutaneous fat tissue. Excise the lymph node before removing the fat pad

6. Transfer the explants from the scale into one well of a 12-well plate prefilled with 1 mL of explant culture medium.
7. Continue massing and transferring explants until the desired number of wells is filled.
8. Maintain explants in a 37 °C incubator with 10 % CO_2 for 2–3 days before co-culturing with hepatocytes (*see* **Note 14**).

3.3 Primary Hepatocyte Isolation

1. Warm the perfusion buffer, the liver digest buffer, and an aliquot of PBS to 42 °C in a water bath placed near the surgical area. Pour 10 mL sterile PBS and 10 mL cold seeding medium into a sterile dish and place on ice (*see* **Note 15**).
2. Anesthetize the first mouse, and after a few minutes squeeze the paws with forceps to ensure deep anesthesia.
3. Turn the pump on to allow warm perfusion buffer to flow through the tubing before beginning surgery.
4. Secure the mouse on the surgical board and cut through both the skin and the peritoneum to open the abdomen. Using forceps, gently push the intestines away from the liver to reveal the inferior vena cava (IVC) and the hepatic portal vein (Fig. 2).
5. Cannulate the IVC with the butterfly needle as buffer is flowing, then immediately cut the hepatic portal vein (*see* **Note 16**).
6. Stabilize the base of the needle with pins. Blood should rapidly clear from the liver, and the color should turn to beige within 10–15 s.
7. After 3 min, quickly turn the pump off, switch to digestion buffer, and restart the pump for an additional 7 min. Periodically pipette warm PBS over the liver (*see* **Note 17**).

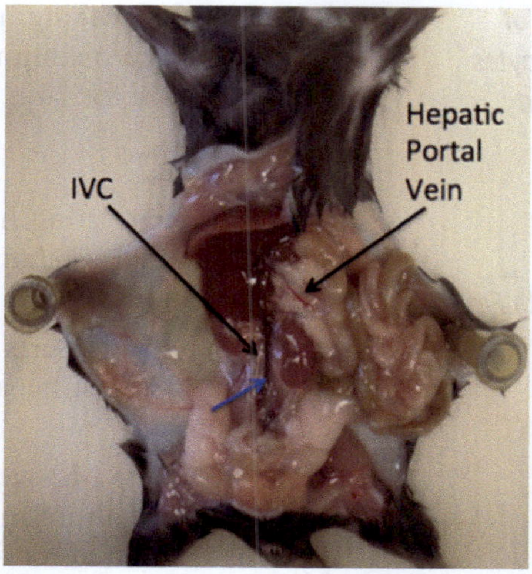

Fig. 2 Surgical access to the inferior vena cava (IVC) and portal vein for hepatocyte isolation. Begin cannulation at the base of the IVC, as shown by the *blue arrow*

8. Carefully cut out the liver, avoiding the gallbladder, and place into the prepared dish. Move to a tissue culture hood to ensure sterile conditions from this point forward.
9. Using two forceps, gently tear apart the liver tissue releasing hepatocytes into solution (*see* **Note 18**).
10. Pour contents over a 100 μm filter into a 50 mL falcon tube, then wash the dish with an additional 10 mL of seeding medium, and pour through the filter.
11. Centrifuge at 150 × *g* for 3 min, then aspirate off the supernatant (*see* **Note 19**).
12. Add 15 mL of seeding medium, gently resuspend cells, and centrifuge again at 150 × *g* for 3 min, then aspirate off the supernatant.
13. Resuspend cells in 10 mL of seeding medium, then add 10 mL of Percoll solution and mix before centrifuging at 250 × *g* for 6 min (*see* **Note 20**).
14. Aspirate off the supernatant, including dead cells and debris floating at the top of the Percoll solution. Resuspend cells in 15 mL of seeding medium and centrifuge again at 150 × *g*.
15. Aspirate off the supernatant, resuspend cells, and count cell number with trypan blue to determine viability and yield.
16. Plate cells at approximately 400,000 cells/mL. For conditioned medium experiments, we typically plate in 24-well plates with 200,000 cells per well, and for explant experiments we plate in 12-well plates with 400,000 cells per well. Maintain cells in a 37 °C incubator with 5 % CO_2.
17. After 4–5 h, aspirate medium and replace with warm culture medium.

3.4 Treatment of Primary Hepatocytes with Conditioned Medium

1. 24–48 h after seeding hepatocytes, wash once with PBS, then replace culture medium with conditioned medium from adipocyte culture for 24 h (*see* **Note 21**).

3.5 Treatment of Primary Hepatocytes with Fat Explants

1. 24–48 h after plating hepatocytes, and 48–72 h after preparing fat explants, aspirate hepatocyte medium and replace with 700 μl fresh hepatocyte culture medium (*see* **Note 22**).
2. Place co-culture inserts into the wells containing hepatocytes, and add 600 μl of hepatocyte culture medium to each insert (*see* **Note 23**).
3. Fill each well of a 12-well plate with PBS, and transfer explants from explant culture medium to PBS to wash, working well-by-well. Then transfer from PBS to the co-culture inserts.
4. Maintain co-culture in a 37 °C incubator with 5 % CO_2 for 24 h.

4 Notes

1. Standard culture of hepatocytes works best with a low level of insulin present. However, if downstream assays require insulin-free conditions, omit insulin from the collection medium.
2. The following protocol refers to explant preparation from inguinal fat; however, it should be applicable to all fat depots. In our experience, interscapular brown fat explants are viable for not more than 3 days in culture, inguinal fat explants for not more than 7 days in culture, and epididymal fat explants for not more than 12 days in culture.
3. The size of the fat depots of the mice determines the number necessary for each experiment. The inguinal depots from one 12-week C57BL/6 male mouse should provide enough tissue for 3–4 wells of a 12-well plate.
4. Scale should be wiped with 70 % ethanol and placed under the tissue culture hood for measuring explant mass.
5. Several companies sell co-culture inserts. We typically use Corning brand inserts with a 0.4 μm pore size to allow diffusion of large macromolecules.
6. The liver of one mouse should yield enough cells for several plates, so typically for conditioned medium and explant experiments one mouse is sufficient.
7. We prefer to use Invitrogen's prepared liver digest medium, which requires a pH adjustment to 7.4.
8. If using Invitrogen's prepared liver digest medium, aliquot the pH adjusted medium into 50 mL aliquots, and freeze in a location that is protected from light. Before each usage, thaw

one aliquot overnight at 4 °C and then filter sterilize. The freeze-thaw cycle can result in precipitates that will occlude small blood vessels if not filtered prior to usage. For this reason, it is important to filter *each* aliquot after thawing.

9. Bubbles can easily form in the warmed buffers and will become lodged in the small blood vessels of the liver.
10. We use a solution of ketamine/xylazine with an IP injection of approximately 175 mg/kg body weight ketamine and 16 mg/kg body weight xylazine.
11. Hepatocytes will adhere to regular, noncoated plates, but collagen coated plates are preferred.
12. Be sure to pipette accurately when adding collection medium. We recommend using micropipettes instead of serological pipettes to ensure equal volumes in each well.
13. If a large number of explants will be prepared, use explant culture medium for this step instead of PBS to maintain viability over the duration of explant preparation and massing.
14. When first isolated, inflammatory genes are significantly upregulated in explants. Inflammation should decrease over the first 48 h in culture, although it will likely never fully return to in vivo levels.
15. Keep the dish near the surgical area so that once the liver is removed it can be immediately placed on ice. Additionally, have a container of ice available in the tissue culture hood so the liver can be transported quickly.
16. When cannulating, begin at the base of the IVC and move the needle at a very shallow angle so it is almost parallel to the IVC. After piercing through the IVC wall, continue to push the tip of the needle up toward the liver before securing the base with surgical pins. Alternatively, the IVC can be cannulated before turning on the pump; however, there is a small danger of bubbles forming in the tip of the needle.
17. If perfusion is done correctly, after 4 or 5 min of perfusion with the digestion buffer the liver should begin to swell and take on a light pinkish hue.
18. It should take a few minutes to dissociate all of the tissue. Avoid aggressively disrupting the liver as hepatocytes are large and fragile cells.
19. After the first centrifugation, the supernatant will be cloudy with debris. After each following centrifugation, the supernatant should be clear of any visible debris.
20. The Percoll solution is a dense mixture that allows living hepatocytes to form a pellet at the bottom, while dead cells will rise to the top of the solution. This separation step is crucial to avoid a low viability of isolated cells.

21. Hepatocytes should live in vitro for >7 days; however, it is recommended to begin experiments within 48 h of plating while cells maintain their cuboidal morphology. Various lengths of treatment with conditioned medium can be used depending on the experimenter's objectives.
22. Omit insulin from the culture medium if future experiments require insulin-free conditions.
23. The amount of medium can be varied as long as the explants are completely submersed. We find that 700 μl in the base of each well and 600 μl added to the inserts is sufficient.

References

1. Rosen ED, Spiegelman BM (2014) What we talk about when we talk about fat. Cell 156: 20–44
2. Zhang Y, Proenca R, Maffei M et al (1994) Positional cloning of the mouse obese gene and its human homologue. Nature 372:425–432
3. Cook KS, Min HY, Johnson D et al (1987) Adipsin: a circulating serine protease homolog secreted by adipose tissue and sciatic nerve. Science 237:402–405
4. Scherer PE, Williams S, Fogliano M et al (1995) A novel serum protein similar to C1q, produced exclusivelt in adipocytes. J Biol Chem 270: 26746–26749
5. Steppan CM, Bailey ST, Bhat S et al (2001) The hormone resistin links obesity to diabetes. Nature 409:307–312
6. Yang Q, Graham TE, Mody N et al (2005) Serum retinol binding protein 4 contributes to insulin resistance in obesity and type 2 diabetes. Nature 436:356–362
7. Dahlman I, Elsen M, Tennagels N et al (2012) Functional annotation of the human fat cell secretome. Arch Physiol Biochem 118:84–91
8. Roca-Rivada A, Belen Bravo S, Perez-Sotelo D et al (2015) CILAIR-based secretome analysis of obese visceral and subcutaneous adipose tissue reveals distinctive ECM remodeling and inflammation mediators. Sci Rep. doi:10.1038/srep12214
9. Cohen P, Levy JD, Zhang Y et al (2014) Ablation of PRDM16 and beige adipose causes metabolic dysfunction and a subcutaneous to visceral fat switch. Cell 156:304–316

Chapter 16

Identification and Quantification of Human Brown Adipose Tissue

Maria Chondronikola, Craig Porter, John O. Ogunbileje, and Labros S. Sidossis

Abstract

Brown adipose tissue (BAT) has attracted significant interest as a potential target tissue against obesity and its associated metabolic perturbations. The purpose of this chapter is to provide an overview of some of the methodological approaches that can be used to identify and quantify BAT in people. Specifically, we will provide a step-by-step description of the following procedures: quantification of BAT in vivo using positron emission tomography-computed tomography (PET/CT) with 2-deoxy-2-[^{18}F]fluoroglucose (^{18}F-FDG) as a tracer, mitochondrial respiration, and uncoupling protein 1 (UCP1) gene and protein expression.

Key words Brown adipose tissue, Positron emission tomography-computed tomography, Uncoupling protein 1, 2-deoxy-2-[^{18}F]fluoroglucose, Biopsy

1 Introduction

Brown adipose tissue (BAT), a recently identified tissue in adults [1–3], has attracted significant interest as a potential target tissue against obesity and its associated metabolic perturbations [4]. Therefore, the methods for the quantification and identification of BAT in people are still in their infancy. Cold exposure in conjunction with PET/CT scans using ^{18}F-FDG as a tracer is a widely used approach for the detection and the quantification of BAT in people. The specific method for the detection of BAT in humans is based on the fact that activated brown adipocytes (via cold or other stimuli) exhibit increased glucose uptake [5]. Upon its uptake by the cells, ^{18}F-FDG is phosphorylated to its 6-phosphate, which cannot be further metabolized and is trapped in the cells. The PET component of the PET/CT detects the radioactivity emitted by the decay of ^{18}F-FDG by each tissue, while the CT measures the density of each tissue. Fused PET/CT images are used to

Jun Wu (ed.), *Thermogenic Fat: Methods and Protocols*, Methods in Molecular Biology, vol. 1566,
DOI 10.1007/978-1-4939-6820-6_16, © Springer Science+Business Media LLC 2017

co-localize areas with increased metabolic activity and density that correspond to adipose tissue, allowing the visualization and quantification of human BAT. Nevertheless, ^{18}F-FDG-PET/CT may yield false-negative results, especially if BAT has not been optimally activated [6], while PET/CT imaging may not be available or permitted in specific populations (children, pregnant or lactating women, patients who already had received large amounts of radiation) or be sensitive enough to detect small changes in BAT, especially those not pertaining to the metabolism of glucose. Therefore, further functional and molecular analysis of BAT samples can provide important information regarding the metabolic function of BAT in people. In this chapter, we describe a comprehensive methodological approach for the identification and quantification of human BAT.

2 Materials

2.1 Cold Exposure Equipment

1. Temperature-controlled room.
2. Liquid-conditioned garments with a temperature-controlled water tank (*see* **Note 1**).

2.2 Imaging and Quantification of Human BAT In Vivo

1. 75–185 Mbq of ^{18}F-FDG.
2. Intravenous catheter (for the administration of ^{18}F-FDG).
3. PET/CT scanner.
4. Fusion software for the quantification of the BAT.

2.3 Sampling

1. Computed tomography (CT) equipment.
2. 25-gauge injection needle.
3. Sterile drapes.
4. Sterile scissors.
5. Sterile scalpel.
6. Lidocaine (2 %).
7. Sterile 6 mm Bergström needle.
8. 60 mL syringe.
9. 10 mL syringe.
10. Suction catheter.
11. Standard 2.0 silk sutures (optional).
12. Antiseptic solution (iodine, chlorexidine).
13. Bacitracin ointment.
14. Sterile garments (gown and gloves).
15. Saline for washing the biopsies.
16. Absorbent material for blotting the tissue.

2.4 BAT Functional and Molecular Analysis In Vitro

2.4.1 Mitochondrial Respiration

1. Oxygraph-2K high-resolution respirometer (Oroboros Instruments, Innsbruck, Austria, *see* **Note 2**).
2. dd H_2O.
3. Ethanol.
4. Micro-pipettes (allowing 10–10,000 μl to be pipetted accurately).
5. Plastic and glass tubes (ranging 500 μl–5 mL in capacity).
6. Weigh boats, petri dishes, and/or 6-well plates (*see* **Note 3**).
7. Fine forceps, sharp scissors, and scalpels (*see* **Note 4**).
8. Stereology microscope (*see* **Note 5**).
9. Filter paper and parafilm (*see* **Note 6**).
10. Microbalance.
11. Biopsy preservation buffer*:* 10 mM CaK_2-EGTA, 7.23 mM K_2-EGTA, 20 mM imidazole, 20 mM taurine, 50 mM K-MES, 0.5 mM dithiothreitol, 6.56 mM $MgCl_2$, 5.77 mM ATP, and 15 mM phosph-creatine (*see* **Note 7**).
12. Respiration buffer*:* 0.5 mM EGTA, 3 mM $MgCl_2$, 60 mM K-lactobionate, 20 mM taurine, 10 mM KH_2PO_4, 20 mM HEPES, 110 mM sucrose, and 1 mg/mL essential fatty acid free bovine serum albumin (*see* **Note 8**).
13. Mitochondrial substrates, inhibitors, and uncouplers: 5 mM Na2-Pyruvate, 1.5 mM octanoyl-l-carnitine, 2 mM malate, 10 mM glutamate, 10 mM succinate, 5 mM K-ADP; 2 μM oligomycin, 20 mM GDP, 10 μM cytochrome C, 5 μM CCCP (*see* **Note 9**).

2.4.2 UCP1 Gene Expression

RNA Extraction Materials

1. RNase ZapTM.
2. 2–mercaptoethanol.
3. Absolute ethanol (96–100 %).
4. RNase-Free Water.
5. Microcentrifuge capable of centrifuging 20,000 × *g*.
6. 5 mL RNase-free microcentrifuge tubes.
7. Phosphate buffer saline (PBS).
8. RNase-free pipet tips.
9. Rotor-stator homogenizer.
10. Micro-volume plate reader.
11. Spin-column-based RNA extraction method.
12. 50–150 mg of adipose tissue.

Conversion of RNA to cDNA Materials

1. Vortexer.
2. Microcentrifuge.
3. Centrifuge with 96-well adapter.
4. 96-well thermal cycler.
5. High-Capacity RNA-to-cDNA Kit.

Gene Expression Materials

1. StepOne Plus real-time PCR system.
2. RNase-free pipet tips.
3. RNase-Free Water.
4. TaqMan® Gene Expression Master Mix.
5. TaqMan® VIC-GAPDH.
6. TaqMan® UCP1 target genes tagged with FAM probe.
7. Optical adhesive film.
8. Optical 96-Well Reaction Plate.
9. 200 ng cDNA samples.

2.4.3 Expression of UCP1 Protein

Adipose Tissue Lysate Preparation

1. T-PER™ Tissue Protein Extraction Reagent.
2. Homogenizer.
3. 50–150 mg of adipose tissue.
4. 1 mL pipette.
5. Bradford Dye Reagent.

Sample Preparation, Sodium Dodecyl Sulfate Polyacrylamide Gel Electrophoresis (SDS-PAGE), Western Blotting

1. Sample Reducing Agent (10×).
2. LDS Sample Buffer (4×).
3. To enable easy analysis, sample buffer and reducing reagent solution is prepared by diluting Sample Reducing Agent (10×) (250 μL) to 4× using 375 μL PBS, bringing the total volume to 625 μL. Add 625 μL LDS Sample Buffer (4×) to the 4× reducing agent, this brings the final concentration to 2× Sample buffer and reducing reagent which can be used with the sample to arrive at 1× sample solution.
4. Antioxidant.
5. Prestained protein standard.
6. Phosphate buffer saline.
7. Ultrapure deionized water.

8. Heat block set at 70 °C.
9. 10× Tris-buffered saline; 24 g Tris–HCl, 5.6 g Tris–base, 88 g of NaCl weighed into deionized water.
10. 4–12 % Bis–tris gel.
11. Transfer Buffer (20×).
12. Mini Gel Tank electrophoresis system and Mini Blot Module.
13. 2-(*N*-morpholino) ethanesulfonic acid–sodium dodecyl sulfate (MES-SDS) running buffer.
14. Immunological detection: 10× TBS, blocking solution (1× TBS, 0.1 % Tween 20, 5 % nonfat dry milk powder), washing solution (1× TBS, 0.1 % Tween 20), enhanced chemiluminescence (ECL) kit, anti-UCP1, and secondary antibody horseradish-peroxidase-labeled (HRP) conjugate.

3 Methods

3.1 BAT Quantification In Vivo

3.1.1 BAT Activation (See Note 10)

1. Expose the participant to mild cold for at least 60 min under fasting conditions trying to avoid/minimize shivering. The cold exposure can be achieved via decreased ambient environmental temperature and liquid conditioned garments (*see* **Note 11**).

3.1.2 BAT Imaging

1. Administer a bolus injection of ^{18}F-FDG via a catheter placed in an antecubital vein in the participant's arm.
2. Ask the participant to remain for approximately 60 min in a dark room with minimal movement under cold exposure covered with liquid conditioned garment.
3. Remove the liquid conditioned garment, if needed.
4. Ask the participant to void his/her bladder.
5. Transfer the participant to the PET/CT scanner location and ask him to lay in a supine position on the moving bed (*see* **Note 12**).
6. Begin the scanning process. The duration of the imaging is approximately 30 min.

3.1.3 BAT Quantification

1. Assess the ^{18}F-FDG PET/CT images (Figs. 1 and 2a) for BAT with a commercially available fusion software using the following criteria:
   (a) ^{18}F-FDG uptake evident in the cervical/supraclavicular, mediastinal, paravertebral, and/or perirenal areas.
   (b) ^{18}F-FDG uptake with a mean SUV of 1.5 or greater (an indicator of ^{18}F-FDG uptake intensity) (*see* **Note 13**).
   (c) The tissue corresponding to the density of adipose tissue on CT (−190 to −30 Hounsfield units) (*see* **Note 13**).
2. Quantify the volume of ^{18}F-FDG BAT by autocontouring each identified individual BAT deposit.

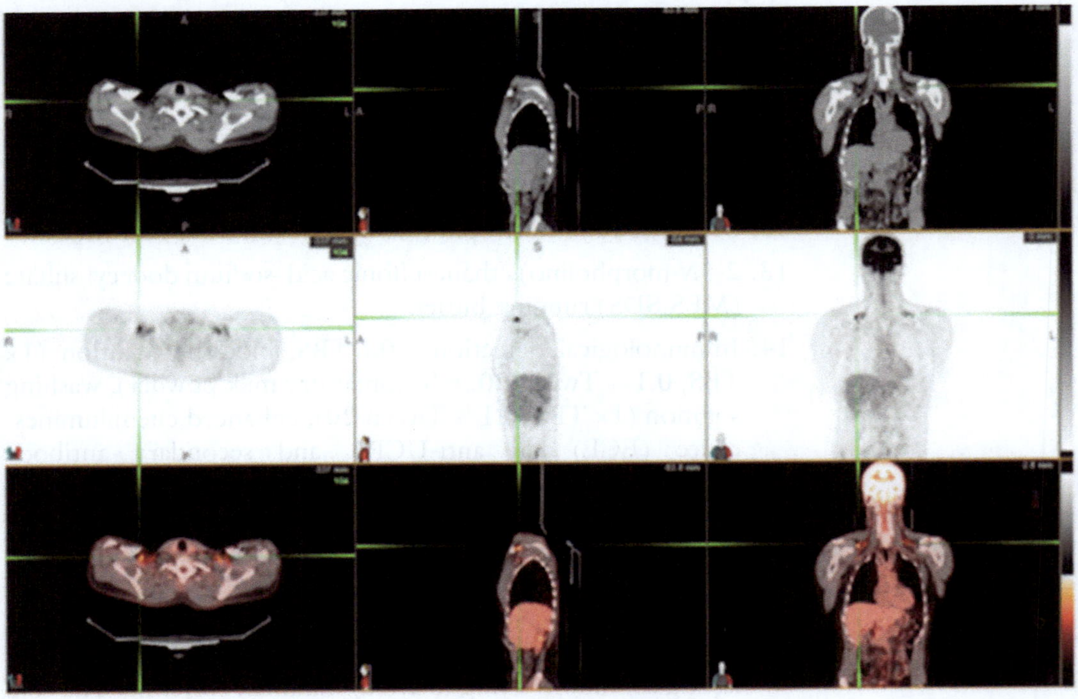

Fig. 1 2-deoxy-2-[^{18}F]fluoro-D-glucose (^{18}F-FDG)-Positron Emission Tomography (PET)/Computed Tomography (CT) imaging for the identification and quantification of human brown adipose tissue (BAT). The *upper row* shows axial, sagittal, and coronal CT images from a representative research participant. The *middle row* shows axial, sagittal, and coronal PET images. The *lower row* presents fused PET/CT images from the same participant. The symmetrical spots with *intense orange color* in the supraclavicular area correspond to metabolically active brown adipose tissue

3.2 Identification and Function and Molecular Assessment of BAT In Vitro

3.2.1 Sampling the Supraclavicular BAT

A properly trained medical professional (surgeon or interventional radiologist) should perform the biopsy procedure [7].

1. Transfer the participant to the CT scanner moving bed and ask him/her to lay down in a supine position (Fig. 2b).
2. Obtain scout images of the lower neck to visualize the anatomy of the supraclavicular area.
3. Identify the location of the supraclavicular BAT depot in the CT images (Fig. 2a) corresponding to the BAT in the PET/CT scan using major anatomical landmarks.
4. Using an antiseptic solution (iodine, chlorexidine, or others), clean thoroughly the skin over the area of the biopsy and cover with a sterile drape.
5. Inject 2–5 mL of 2 % lidocaine in the dermis of the biopsy site and advance the injection needle in the targeted adipose tissue area.
6. Obtain additional scout images of the supraclavicular area to ensure the correct placement of the needle at the biopsy cite; avoid any adjacent superficial veins; and determine the depth of the BAT deposit and the advancement of the biopsy needle (Fig. 2c).

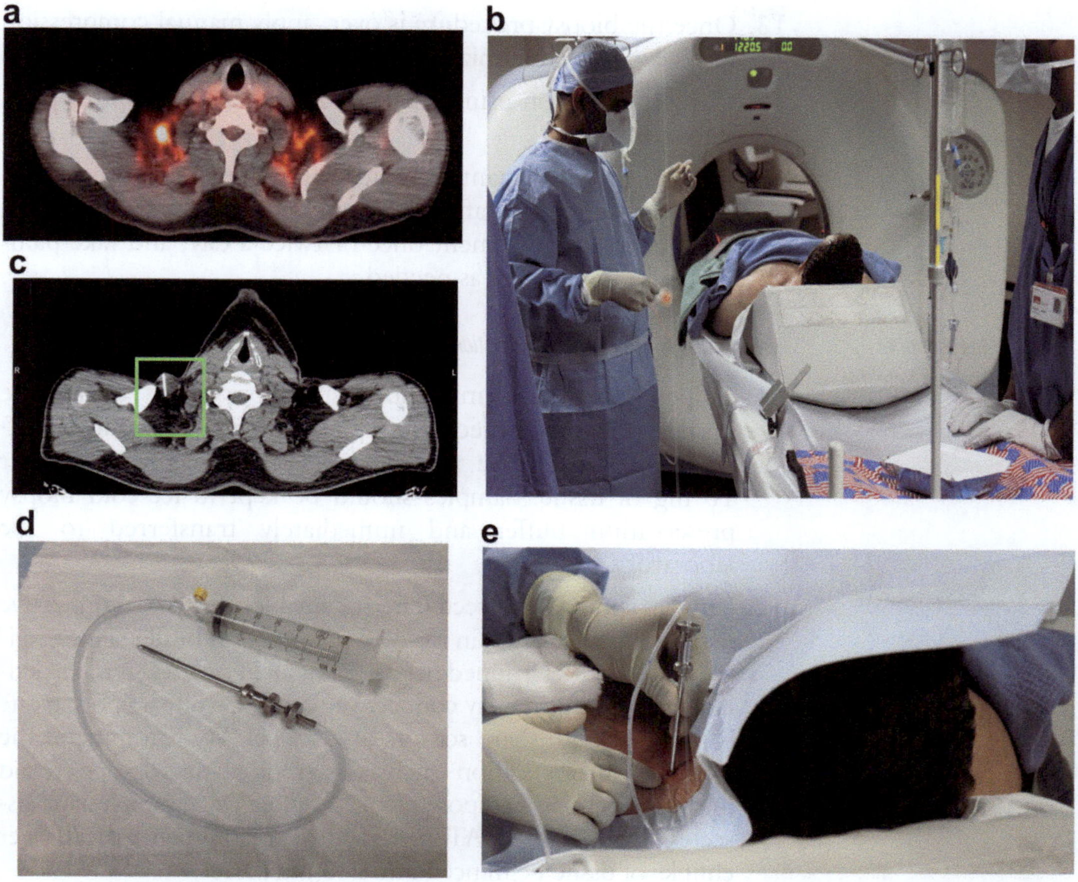

Fig. 2 Brown adipose tissue biopsy. (**a**) 2-deoxy-2-[18F]fluoro-D-glucose (^{18}F-FDG)- Positron Emission Tomography/Computed Tomography image from a study participant. The *intense orange color* corresponds to metabolically active brown adipose tissue. (**b**) The study participant lying on the table of the computed tomography (CT) procedure room. (**c**) CT scout image corresponding to the supraclavicular area after the insertion of the injection needle to ensure its placement. (**d**) Tubing attached to the 6 mm Bergstrom needle and 60 mL BD syringe to generate suction. (**e**) The study physician during the biopsy procedure. Reproduced from Chondronikola and Annamalai et al. [7] with permission from Nature Publishing Group & Palgrave Macmillan

7. Using a scalpel, make an incision (<1 cm) in the skin.
8. Connect a 6 mm Bergström with a suction catheter using a 60 mL syringe (Fig. 2d).
9. Apply precooled bacitracin ointment to the top of the inner cannula to maintain suction and avoid leakage of air between the needle and the inner cannula.
10. Insert the 6 mm Bergström needle through the incision and advance it into the supraclavicular adipose tissue depot (Fig. 2e).
11. Once the needle is positioned appropriately, apply suction using the 60 mL syringe.
12. To sample the supraclavicular adipose tissue, rotate the needle by 180°.

13. Once the biopsy procedure is over, apply manual compression for 3–5 min to minimize bleeding.
14. Close the incision using standard 2.0 silk sutures and apply a sterile dressing.
15. Instruct the participant to avoid swimming and rigorous exercise for the first 72 h after the biopsy; inspect the wound daily; apply antibiotic ointment once or twice a day; and take painkillers for discomfort as needed.

3.2.2 Mitochondrial Respiration (See Note 14)

Preparation of the Tissue for Functional Analysis

1. Adipose tissue for mitochondrial respirometry should be immediately submerged in an ice-cold biopsy preservation buffer (pH of 7.1). We recommend approximately 1 mL per 10 mg of tissue. Samples should be kept in ice-cold biopsy preservation buffer and immediately transferred to the laboratory.
2. Samples should be dissected free of any blood and/or connective tissue prior to analysis in ice-cold preservation buffer since respiration will be determined per mg of tissue. For BAT, it is particularly important that any connective tissue or white adipose tissue is removed with sharp scissors or a scalpel. We recommend the use of low magnification (such as a stereology microscope) to aid in the dissection of adipose tissue samples. Once samples are dissected free of non-BAT material, we recommend that larger chunks of tissue be minced into 2–5 mg pieces.
3. The dissection described above should be performed while samples are bathed in ice-cold preservation buffer. We recommend placing samples and buffer in the well of a 6-well plate, or in a petri-dish or weigh boat, then placing them on ice. If samples are placed under a stereology microscope for prolonged periods of time, we recommend placing the plate/dish containing the sample and buffer on a cold surface. This can be achieved by fashioning a small ceramic or metal block that can be pre-cooled in a refrigerator/freezer.
4. Adipose tissue samples should be weighed on a precision micro-balance prior to analysis. To do this, we recommend briefly blotting samples using filter paper to absorb excess preservation buffer and transferring sample onto a tarred balance. As soon as the sample weight is recorded, samples can be returned to chilled preservation buffer. We find that 5–10 mg wet weight is sufficient for respirometry assays of human BAT, whereas 50–100 mg wet weight is desirable for respirometry assays of human WAT.
5. The preparation of human adipose tissue as described above takes less than 1 h. Thus, it is feasible that respirometry assays

can begin 1–2 h after sample collection. However, based on our experience, human adipose tissue samples can be successfully preserved in ice-cold biopsy preservation buffer for up to 24 h. With that said, we recommend performing respiration assays as soon as possible after the time of sample collection.

High-Resolution Respirometry

1. All polygraphic oxygen sensors should be calibrated daily in respiration buffer prior to sample analysis. Further, periodic instrumental background and zero O_2 calibrations should be performed to insure instrument stability. This is critical for the production of quantitative data when samples are collective from humans prospectively over several months/years. Numerous vendors supply tissue Clarke-type electrode respirometer, and similar instrumental background calibrations can be performed on these instruments to ensure stability over time. In our laboratory, we have used the dual chamber Oxygraph-O2K high-resolution respirometer (Oroboros Instruments, Innsbruck, Austria). See notes for details on how to perform calibrations on the Oroboros Instruments Oxygraph-O2K (*see* **Note 15**).
2. We recommend that high-resolution respirometry measurements in permeabilized human adipose tissue be performed in a respiration buffer (pH 7.1). Adipose tissue can be placed directly into the respirometer chamber using forceps before closing the chamber. The magnetic stir bar of the Oxygraph-2K respirometer should be set at 750 rpm to insure adequate mixing of the respiration buffer. The temperature should be set at 37 °C for all experiments.
3. Several different titration protocols of substrates, inhibitors, and uncouplers can be performed to comprehensively assay mitochondrial respiratory capacity and coupling control. Critically, to directly assay UCP1 function, the titration of substrates (i.e., pyruvate, octanoyl-l-carnitine, malate, glutamate) should be followed by the titration of saturating levels of GDP. Since GDP inhibits UCP1, this protocol allows determination of total mitochondrial leak respiration and UCP1-dependent leak respiration.
4. It should be noted that, while titration of the ATP synthase inhibitor oligomycin allows the determination of leak respiratory capacity, this reflects the total leak respiratory capacity of the sample and is not a measure of UCP1 function. Similarly, titration of uncouplers like CCCP provides a measure of total respiratory capacity of the sample; this also is not a measure of UCP1 function. While titration of both oligomycin and CCCP can provide important information on mitochondrial respiratory capacity and coupling control, these parameters are second to direct measurement of UCP1 function by titration of GDP.

3.2.3 Assessment of Expression of UCP1 Gene Expression in Human BAT

RNA Extraction [8]

1. Prepare fresh lysis buffer (provided with the extraction kit) containing 1 % 2-mercaptoethanol under the fume hood (*see* **Note 16**).
2. Weigh 100–200 mg of fast frozen adipose tissue into a 5 mL microcentrifuge tube and add 300–600 μL prepared lysis buffer.
3. Homogenize using rotor-stator homogenizer until fat is completely homogenized.
4. Vortex until homogenized fat tissue appeared completely lysed.
5. Then centrifuge the homogenate at ~3000 × *g* for 5 min at room temperature. Transfer the clear supernatant into a 5 mL RNase-free microcentrifuge tube (*see* **Note 17**).
6. Add equal volume of the prepared 70 % ethanol in RNase-free water to each fat homogenate and vortex vigorously to ensure proper dispersion of any visible precipitate formed in the homogenate due to the addition of ethanol (*see* **Note 18**).
7. Transfer 700 μL of the homogenate to the spin cartridge placed in the collecting tube. Centrifuge for 15 s at 20,000 × *g* at room temperature. Discard the flow-through and the tube; reinsert the spin cartridge into a new collecting tube (PCR microcentrifuge tube can be used). Repeat this process until the entire sample homogenate has been processed.
8. Add 700 μL of wash buffer I to the spin cartridge containing the extract and centrifuge at 20,000 × *g* for 15 s at room temperature. Place the spin cartridge into a new collecting tube after discarding the flow-through.
9. Add 500 μL of wash buffer II (containing 16 mL of absolute ethanol) to the spin centrifuge and centrifuge for 15 s at room temperature. Discard the flow-through. Repeat this step once.
10. To ensure that all liquids are eliminated from the spin cartridge and to dry the membrane with the attached RNA, centrifuge the spin cartridge at 20,000 × *g* for 2 min. Discard the collecting tube and place the spin cartridge into the recovery tube.
11. To the center of the spin cartridge, add 30–100 μL of RNase-free water. In our laboratory, 30–50 μL of RNase-free water is usually added because the RNA concentration in adipose tissue is limited. Incubate the setup for 1 min at room temperature to help dissolve the RNA.
12. Centrifuge the Spin Cartridge for 2 min at 20,000 × *g* at room temperature to elute the RNA from the membrane into the recovery tube.
13. Determine the recovered RNA concentration and 260/280 ratio. Our laboratory uses RNA with 260/280 ratio of 1.8 or above.

14. Preserve the purified RNA at −80 °C or proceed with cDNA synthesis.

RNA to cDNA Transcription [9]

1. Use approximately 300 ng of purified RNA per 20 μL reaction solution for the transcription of RNA to cDNA using a High Capacity RNA-to-cDNA kit. Thaw kit on ice before experiment.
2. Prepare reaction solution volume enough for the number of samples. To prepare 20 μL reaction's solution, please *see* Table 1. Excess volume should be prepared to allow for volume lost during pipetting. Mix reaction mixture gently on ice and centrifuge at low speed after sealing the reaction tubes/plates.
3. Place the tubes in a 96-well thermal cycler and program; Step 1 at 37 °C for 60 min, Step 2 at 95 °C for 5 min, and Step 3 at 4 °C for ∞. After programing the thermal cycler, start the RNA transcription to cDNA. The produced cDNA can be stored at −20 °C.

Gene Expression

Our laboratory uses multiplex methods for TaqMan gene expression analysis. The house keeping gene is tagged with VIC fluorescent dye while the target genes are tagged with FAM/MGB probe.

1. Allow TaqMan UCP1 gene and TaqMan® Gene Expression Master Mix to thaw on ice.
2. Prepare assay master mixture enough for 20 μL reaction per well using Table 2. It is also important to prepare excess volume to help correct for loss due to pipetting.

Table 1
Preparation of reverse transcriptase (RT) reaction mix using high capacity RNA-to-cDNA kit

Component	Volume/Reaction (μL)	
	+RT	−RT
2× RT buffer	10.0	10.0
20× RT enzyme mix	1.0	
Nuclease-free H_2O	Enough to make up to 20 μL	Enough to make up to 20 μL
Sample	Up to 9 μL	Up to 9 μL
Total per reaction	20.0	20.0

−RT served as control of the transcription reaction

Table 2
Real-time polymerase chain reaction (RT-PCR) sample reaction mix preparation protocol

Component	1 Sample	52 Samples
TaqMan Gene Master Mix	10	520
UCP1-FAM (20×)	1	52
GAPDH-VIC (60×)	0.333333	17.33333
RNase-free water	6.666667	346.6667
Total assay mixture (μL)	18	936

Final sample reaction mix volume is brought up to 20 μL with the addition of 2 μL cDNA, while 2 μL of RNAase-free water is added to the non-template control

3. Briefly vortex assay master mix for 1–2 s.
4. Dispense 18 μL assay master mix to each well of the MicroAmpTM Fast Optical 96-Well Reaction Plate.
5. After dispensing the assay master mix, add 2 μL of cDNA samples to the appropriate wells except for the non-template control (NTC) wells.
6. For the NTC wells, add 2 μL RNase-free water to bring the total volume of the reaction mixture to 20 μL.
7. Seal the wells carefully with MicroAmpTM optical adhesive film.
8. Gently spin the reaction mixture plate at 800 rpm for ~30 s.
9. Program StepOne Plus real-time PCR system for multiplex reaction for 40 cycles.
10. Place reaction mixture plate into the PCR machine and start the amplification reaction.
11. Determine delta-delta CT values to derive thee relative gene expression and fold difference.

3.2.4 UCP1 Protein Expression

Sample Preparation

1. Weigh 50–200 mg of fast frozen adipose tissue into a pyrogen-free vial.
2. Add 150–300 μL of T-PER™ lysis buffer containing protease-phosphatase inhibitor cocktail at the ratio of 5 μg–200 μL of buffer. Homogenize tissue using dounce homogenizer or mechanical homogenizer (*see* **Note 19**).
3. Incubate homogenate on ice for 30 min.
4. Spin lysate in a pre-cold (4 °C) centrifuge for 10 min at 1,380 × *g* and collect supernatant into a fresh tube on ice and store frozen at −80 °C (aliquot to prevent freeze and thaw) (*see* **Note 20**).

5. Determine lysates protein concentration using modified Bradford protein assay kit with a spectrophotometer.

SDS-Page, Western Blotting, and Detection

1. Take equal volume of adipose lysate protein (~20 μg) and loading (sample) buffer and reducing reagent. Bring the volume of all samples to have equal final volume with deionized water.
2. Heat lysate in sample buffer at 70 °C for 10 min using heat block and rapidly cool down.
3. Load prestained protein standard in the first well and samples into 4–12 % Tris-Glycine Gel for gel electrophoresis and run for 120 min at 100 V using Mini-gel Tank electrophoresis system in 1× 2-(*N*-morpholino) ethanesulfonic acid–sodium dodecyl sulfate (MES-SDS) Running Buffer.
4. Transfer protein using Mini Blot Module onto a nitrocellulose membrane using 1× transfer buffer containing 20 % methanol at 20 V for 2 h.
5. To detect UCP1, first block membrane using 5 % not fat milk in TBST for 1 h with gentle rocking at room temperature. This will help reduce nonspecific binding of the antibody of interest. After blocking, expose membrane to anti-UCP1 antibody, dilute at 1:1000 in 5 % milk in 0.1 % TBST, and incubate at 4 °C overnight with gentle shaking.
6. Wash blot three times for 15 min each using 0.1 % TBST at room temperature.
7. Dilute secondary antibody (In our lab, we use donkey anti-rabbit HRP conjugate) at 1:10,000 in 5 % milk in 0.1 % TBST and incubate the membrane for 1 h at room temperature. Wash membrane as indicated in **step 6**.
8. Detect protein after incubation of blots using ECL Western blotting Substrate solution for 1 min at room temperature, wrap in transparent membrane, and visualize with autoradiography (X-RAY film) (Fig. 3).

4 Notes

1. Although BAT activation can be achieved just with air-cooling, we recommend using liquid condition garments to minimize the effect of changes in body posture aiming to conserve heat.
2. While there are numerous other Clarke electrode-based instruments that allow the amperometric determination of O_2 concentration (i.e., Hansatech Instruments, Strathkelvin Instruments), or the high-throughput phosphorescent-based XFe analyzer

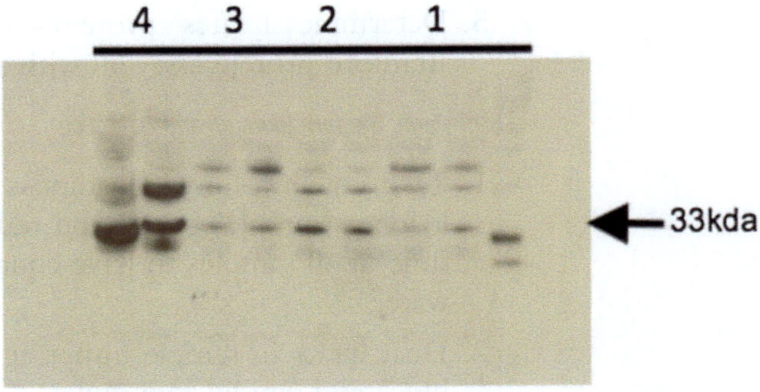

Fig. 3 Western blot pictograph of mouse uncoupling protein 1 (UCP1) at 33 kda from adipose tissue after burn injury probed with rabbit anti-UCP1 antibody. Twenty micrograms of homogenate was used for analyses. Numbers 1, 2, 3, and 4 represent samples from different mice

(Seahorse Bioscience), we have found the Oxygraph-O2K (Oroboros Instruments) to be advantageous for the determination of human BAT O_2 consumption rate. Firstly, the Oxygraph-2K allows for the high-resolution determination of the O_2 consumption rate in permeabilized whole tissue. Second, O_2 concentration and flux are determined in a large (2 mL) volume of respiration buffer in a closed chamber composed of glass and plastic polymers with low O_2 permeability. Third, Datlab software (Oroboros Instruments, Innsbruck, Austria) permits the polygraphic determination of temperature, O_2 concentration, and O_2 flux at 2 s intervals. Importantly, this allows the experimenter to visual the O_2 flux in real time to establish when a steady state has been achieved. Fourth, multiple user-specific mitochondrial substrates and inhibitors can be titrated into the chamber of the Oxygraph-2K, allowing specific mitochondrial functions (i.e., UCP1 function) to be probed. A detailed description of the air calibration of the Oroboros-O2K polygraphic oxygen sensors can be found on the manufacturer's website: http://www.bioblast.at/images/a/ab/MiPNet19.18D_O2k-Calibration.pdf

3. We have found that plastic weigh boats or 6-well micro-plates are ideal for the short-term storage and dissection of tissue samples.
4. The dissection of tissue samples and the mincing of tissue into smaller (~5 mg pieces) can be performed using a variety of different dissection utensils.
5. We find that it is helpful to view samples under low magnification to identify any blood vessels, connective tissue, and other potential contaminants.
6. To accurately weigh adipose tissue samples, we recommend first blotting samples on filter paper, then placing the tissue directly onto parafilm on a microbalance.

7. For detailed information on suggested vendors for chemicals and for the preparation and storage of chemicals and buffers, we refer the reader to the Oroboros Instruments website: http://www.bioblast.at/index.php/Bioblast
8. Information on the preparation and storage of respiration buffer (MIR05 buffer) can be found at the following link: http://www.bioblast.at/images/d/d9/MiPNet14.13_Medium-MiR06.pdf
9. Mitochondrial substrates are required to provide the electron transport chain with reducing equivalents. In our experiments, we tend to use a combination of 5 substrates (pyruvate, octanoyl-l-carnitine, malate, glutamate, and succinate). However, this may be modified for specific experiments, where individual substrates or different substrate combinations can be used to support mitochondrial respiration. ADP titrations may be utilized to demonstrate mitochondrial coupling control (or lack of). Specifically for BAT, titration of the UCP1 inhibitor GDP allows UCP1 function to be directly assayed. The titration of cytochrome C allows investigators to determine whether the outer mitochondrial membrane is intact or not. Finally, the titration of an uncoupler such as carbonyl cyanide m-chlorophenyl hydrazine (CCCP) allows mitochondrial respiration to be determined in the uncoupled state.
10. Currently, prolonged cold exposure is the method most commonly used to activate BAT in humans. Although beta agonists have been also used also as a means to activate BAT, the administered doses were not adequate to safely stimulate increased BAT metabolic activity [10, 11] or have been tested only in young healthy individuals [12]. Different studies have used different durations of cold exposure to activate BAT, varying from 2 to 6 h [2, 13, 14]. Unfortunately, the protocols that have been used to activate BAT are quite variable; therefore, they cannot be compared directly. However, results from a recent rodent study [15] along with our preliminary data support that a prolonged cold exposure is more probable to activate BAT in it maximal capacity.
11. Although a specific threshold to activate BAT has not been established, temperatures below 19 °C or close to the participant's individual shivering threshold can activate existing BAT depots in humans.
12. Since the cooling vest and the temperature in the PET/CT room are hard to regulate, participants can be asked to place their feet in ice (or ice water) as tolerated and not remain still.
13. Other teams have used different cutoffs for the determination of BAT volume using ^{18}F-FDG PET/CT imaging. These cutoffs have been discussed in a greater detail elsewhere [16].

14. Adipose tissue mitochondrial respiration can be assayed in isolated organelles [17], in cells isolated from tissue samples [18], or in permeabilized tissue [19–23]. While isolation of cells or mitochondria can be readily performed on necropsy samples of rodent BAT, sufficient tissue volume from human biopsy samples cannot be guaranteed. Further, unlike the homogeneous intrascapular BAT pad of rodents, the heterogeneous nature of human BAT biopsies means that careful dissection of BAT away from surrounding WAT and connective tissue is important prior to biochemical analysis, a process that further reduces the amount of tissue available for respirometry experiments. Therefore, in our experience the use of permeabilized biopsy tissue provides the most viable option for the determination of BAT mitochondrial respiratory capacity and coupling control in tissue sampled from humans. Thus, the subsequent description of materials and methodologies for the determination of mitochondrial respiratory function in human BAT pertains to the measurement of oxygen consumption rate in small (2–10 mg) permeabilized BAT samples. For a detailed description of how to isolate cells or mitochondrial from adipose tissue for respiration analysis, please refer to [24, 25].
15. A detailed description of the instrumental background calibration of the Oroboros-O2K polygraphic oxygen sensors can be found on the website of the manufacturer: http://www.bioblast.at/images/6/65/MiPNet14.06_InstrumentalO2Background.pdf http://www.bioblast.at/images/7/77/MiPNet06.03_POS-Calibration-SOP.pdf. Similar calibrations can be determined on instruments produced by other companies (i.e. using dithionite for zero calibrations). Investigators should refer to the operating manual of their respirometer to devise the most appropriate instrumental background calibration regime.
16. Do not place the lysis buffer on ice and prepare 70 % ethanol in RNase-free water.
17. Ensure that fat droplet is not transferred with the supernatant; this can block the spin cartridges and reduce the RNA recovery.
18. Do not use cold 70 % ethanol; it can form a precipitate with the guanidinium present in the lysis at low temperature. This can result in low RNA yield.
19. All work should be done on ice.
20. Caution should be taken when pipetting lysate not to transfer fat droplet alongside the homogenate. Fat droplet can alter result during electrophoresis. If fat droplet is found in the lysate, it is strongly recommended to respin the lysate.

References

1. Cypess AM, Lehman S, Williams G, Tal I, Rodman D, Goldfine AB, Kuo FC, Palmer EL, Tseng YH, Doria A, Kolodny GM, Kahn CR (2009) Identification and importance of brown adipose tissue in adult humans. N Engl J Med 360(15):1509–1517. doi:10.1056/NEJMoa0810780
2. van Marken Lichtenbelt WD, Vanhommerig JW, Smulders NM, Drossaerts JM, Kemerink GJ, Bouvy ND, Schrauwen P, Teule GJ (2009) Cold-activated brown adipose tissue in healthy men. N Engl J Med 360(15):1500–1508. doi:10.1056/NEJMoa0808718
3. Virtanen KA, Lidell ME, Orava J, Heglind M, Westergren R, Niemi T, Taittonen M, Laine J, Savisto NJ, Enerback S, Nuutila P (2009) Functional brown adipose tissue in healthy adults. N Engl J Med 360(15):1518–1525. doi:10.1056/NEJMoa0808949
4. Porter C, Chondronikola M, Sidossis LS (2015) The therapeutic potential of brown adipocytes in humans. Front Endocrinol 6:156. doi:10.3389/fendo.2015.00156
5. Hany TF, Gharehpapagh E, Kamel EM, Buck A, Himms-Hagen J, von Schulthess GK (2002) Brown adipose tissue: a factor to consider in symmetrical tracer uptake in the neck and upper chest region. Eur J Nucl Med Mol Imaging 29(10):1393–1398. doi:10.1007/s00259-002-0902-6
6. Lee P, Greenfield JR, Ho KK, Fulham MJ (2010) A critical appraisal of the prevalence and metabolic significance of brown adipose tissue in adult humans. Am J Physiol Endocrinol Metab 299(4):E601–E606. doi:10.1152/ajpendo.00298.2010
7. Chondronikola M, Annamalai P, Chao T, Porter C, Saraf MK, Cesani F, Sidossis LS (2015) A percutaneous needle biopsy technique for sampling the supraclavicular brown adipose tissue depot of humans. Int J Obes (Lond). doi:10.1038/ijo.2015.76
8. Applied Biosystems (2012). PureLink RNA mini kit https://tools.thermofisher.com/content/sfs/manuals/purelink_rna_mini_kit_man.pdf
9. Applied Biosystems (2007) High Capacity-RNA-to-cDNA Kit. http://www3.appliedbiosystems.com/cms/groups/mcb_support/documents/generaldocuments/cms_047249.pdf
10. Cypess AM, Chen YC, Sze C, Wang K, English J, Chan O, Holman AR, Tal I, Palmer MR, Kolodny GM, Kahn CR (2012) Cold but not sympathomimetics activates human brown adipose tissue in vivo. Proc Natl Acad Sci U S A 109(25):10001–10005. doi:10.1073/pnas.1207911109
11. Vosselman MJ, van der Lans AA, Brans B, Wierts R, van Baak MA, Schrauwen P, van Marken Lichtenbelt WD (2012) Systemic beta-adrenergic stimulation of thermogenesis is not accompanied by brown adipose tissue activity in humans. Diabetes 61(12):3106–3113. doi:10.2337/db12-0288
12. Cypess AM, Weiner LS, Roberts-Toler C, Elia EF, Kessler SH, Kahn PA, English J, Chatman K, Trauger SA, Doria A, Kolodny GM (2015) Activation of human brown adipose tissue by a beta3-adrenergic receptor agonist. Cell Metab 21(1):33–38. doi:10.1016/j.cmet.2014.12.009
13. Chondronikola M, Volpi E, Borsheim E, Porter C, Annamalai P, Enerback S, Lidell ME, Saraf MK, Labbe SM, Hurren NM, Yfanti C, Chao T, Andersen CR, Cesani F, Hawkins H, Sidossis LS (2014) Brown adipose tissue improves whole-body glucose homeostasis and insulin sensitivity in humans. Diabetes 63(12):4089–4099. doi:10.2337/db14-0746
14. Yoneshiro TAS, Matsushita M, Kameya T, Nakada K, Kawai Y, Saito M (2011) Brown adipose tissue, whole-body energy expenditure, and thermogenesis in healthy adult men. Obesity 19(1):13–16
15. Labbe SM, Caron A, Bakan I, Laplante M, Carpentier AC, Lecomte R, Richard D (2015) In vivo measurement of energy substrate contribution to cold-induced brown adipose tissue thermogenesis. FASEB J 29(5):2046–2058. doi:10.1096/fj.14-266247
16. van der Lans AA, Wierts R, Vosselman MJ, Schrauwen P, Brans B, van Marken Lichtenbelt WD (2014) Cold-activated brown adipose tissue in human adults: methodological issues. Am J Physiol Regul Integr Comp Physiol 307(2):R103–R113. doi:10.1152/ajpregu.00021.2014
17. Shabalina I, Petrovic N, de Jong J, Kalinovich A, Cannon B, Nedergaard J (2013) UCP1 in Brite/Beige adipose tissue mitochondria is functionally thermogenic. Cell Rep 5:1196–1203
18. Matthias A, Ohlson K, Fredriksson J, Jacobsson A, Nedergaard J, Cannon B (2000) Thermogenic responses in brown fat cells are fully UCP1-dependent. J Biol Chem 275:25073–25081
19. Kraunsøe R, Boushel R, Hansen C, Schjerling P, Qvortrup K, Støckel M, Mikines K, Dela F

(2010) Mitochondrial respiration in subcutaneous and visceral adipose tissue from patients with morbid obesity. J Physiol 588:2023–2032

20. Porter C, Herndon D, Bhattarai N, Ogunbileje J, Szczesny B, Szabo C, Toliver-Kinsky T, Sidossis L (2015) Severe burn injury induces thermogenically functional mitochondria in murine white adipose tissue. Shock 44:258–264
21. Sidossis L, Porter C, Saraf M, Borsheim E, Radhakrishnan R, Chao T, Ali A, Chondronikola M, Mlcak R, Finnerty C, Hawkins H, Toliver-Kinsky T, Herndon D (2015) Browning of subcutaneous white adipose tissue in humans after severe adrenergic stress. Cell Metab 22:219–227
22. Chondronikola M, Annamalai P, Chao T, Porter C, Saraf M, Cesani F, Sidossis L (2015) A percutaneous needle biopsy technique for sampling the supraclavicular brown adipose tissue depot of humans. Int J Obes (Lond). Epub ahead of print
23. Chondronikola M, Volpi E, Børsheim E, Porter C, Annamalai P, Enerbäk S, Liddell ME, Saraf M, Labbe S, Hurren S, Yfanti C, Chao T, Andersen C, Cesani F, Hawkins H, Goodwin J, Sidossis L (2014) Brown adipose tissue activation improves glucose homeostasis and insulin sensitivity in humans. Diabetes 63(12):4089–4099
24. Cannon B, Nedergaard J (2001) Studies of thermogenesis and mitochondrial function in adipose tissues. Methods Mol Biol 456:109–121
25. Cannon B, Nedergaard J (2001) Respiratory and thermogenic capacities of cells and mitochondria from brown and white adipose tissue. Methods Mol Biol 155:295–303

Chapter 17

Designing 3-D Adipospheres for Quantitative Metabolic Study

Takeshi Akama, Brendan M. Leung, Joe Labuz, Shuichi Takayama, and Tae-Hwa Chun

Abstract

Quantitative assessment of adipose mitochondrial activity is critical for better understanding of adipose tissue function in obesity and diabetes. While the two-dimensional (2-D) tissue culture method has been sufficient to discover key molecules that regulate adipocyte differentiation and function, the method is insufficient to determine the role of extracellular matrix (ECM) molecules and their modifiers, such as matrix metalloproteinases (MMPs), in regulating adipocyte function in three-dimensional (3-D) in vivo-like microenvironments. By using a 3-D hanging drop tissue culture system, we are able to produce scalable 3-D adipospheres that are suitable for quantitative metabolic study in 3-D microenvironment.

Key words Spheroid, Drop culture, Adipocyte, Mitochondria

1 Introduction

In the adipose tissue, a group of adipocytes are clustered to form a grape-like structure that is enmeshed with collagen fibers [1, 2]. Within this three-dimensional (3-D) ECM scaffold, vascular and autonomic nervous system are intertwined to regulate the metabolic function of the adipose tissue [3]. Vascular system provides nutrition and oxygen necessary for adipose tissue function, whereas autonomic nervous system regulates the lipolysis of adipocytes. Moreover, preadipocytes within adipose tissues may regulate the function of the adipose tissue through de novo adipogenesis as well as the interaction with adipocytes in paracrine and juxtacrine manners. Adipocyte differentiation and function in 3-D environment significantly differ from those in 2-D [1]; therefore, assessing the metabolic function of the adipose tissue in a 3-D configuration is a critical step forward to better understand the 3-D adipose tissue metabolism. Herein, we describe a high-throughput tissue culture method to prepare 3-D adipospheres [4, 5] and measure their mitochondrial respirations. Through this approach, we are able to

Jun Wu (ed.), *Thermogenic Fat: Methods and Protocols*, Methods in Molecular Biology, vol. 1566,
DOI 10.1007/978-1-4939-6820-6_17,

control the quantity of adipocytes per miniature adipose tissues as well as ECM components depending on the experimental purposes. The physiological and pathological (fibrotic) 3-D adipospheres are ideal platforms to determine adipose tissue function in the in vivo-like environment. 3-D adipospheres can serve as a functional adipose unit useful for the better understanding of 3-D adipose tissue function.

2 Materials

2.1 Hanging Drop Culture of Preadipocyte Spheroids

1. Perfecta3D Hanging Drop Plates from 3D Biomatrix, Ann Arbor, MI, USA.
2. Sterilized water.
3. Hanging drop culture medium: high-glucose DMEM containing 10 % fetal bovine serum (FBS), 2 mM glutamine, 100 U/ml of penicillin, and 100 U/ml of streptomycin.
4. 1.2 % Methocel A4M (Sigma, St. Louis, MO, USA), dissolved into hanging drop culture media.
5. Multichannel pipette.
6. Water reservoir.
7. 3T3L1 cells (ATCC, Manassas, VA) or any other preadipocytes.

2.2 Induction of Adipocyte Differentiation

1. Adipocyte differentiation medium: high-glucose DMEM, 10 % FBS, 2 mM glutamine, 100 U/ml of penicillin, 100 U/ml of streptomycin, 250 nM dexamethasone, 10 μM troglitazone, 10 nM T3, and 1 μg/ml insulin.
2. Insulin medium: high-glucose DMEM, 10 % FBS, 2 mM glutamine, 100 U/ml of penicillin, 100 U/ml of streptomycin, and 1 μg/ml insulin.

2.3 Mitochondrial Respiration Assay

1. XFe96 Spheroid FluxPak (Seahorse Biosciences, North Billerica, MA, USA).
2. Corning Costar Ultra-Low attachment multiwell plate (Corning, Corning, NY, USA).
3. Mitochondrial Assay Medium: XF Assay Medium Modified DMEM (Seahorse Biosciences), 25 mM glucose, and 2 mM sodium pyruvate adjusted to pH 7.3–7.4, more than 50 ml (*see* **Note 1**).
4. 3 mL of 2 μM oligomycin (Sigma) in mitochondrial assay medium.
5. 3 mL of 0.5 μM carbonyl cyanide-p-trifluoromethoxyphenylhydrazone (FCCP, from Sigma) in assay medium.

6. 3 mL of 1 μM antimycin A and 1 μM rotenone (all from Sigma) in assay medium.
7. Incubator at 37 °C supplying no CO_2.
8. XFe96 Extracellular flux analyzer with a Spheroid microplate-compatible thermal tray from Seahorse Biosciences.

3 Methods

3.1 Spheroid Formation and Adipocyte Differentiation

1. Supply sterilized water into designated hydration reservoir in a bottom plate or in a bottom tray (*see* **Note 2**).
2. Prepare cell suspension containing appropriate number of cells in 25 μl hanging drop culture medium per drop (Fig. 1, *see* **Note 3**) [5]. The addition of Methocel at 0.24 % final concentration promotes the formation of a single spheroid per hanging drop instead of multiple small satellite spheroids.
3. Transfer cell suspension into a reservoir trough and dispense 25 μl each into a hanging drop plate using a multichannel pipette.
4. Hanging drop plate can be detached from the bottom reservoir and placed directly on a microscope stage during observation. Confirm that a single spheroid is formed in a well after one day. At 2 days post-seeding change media to adipocyte

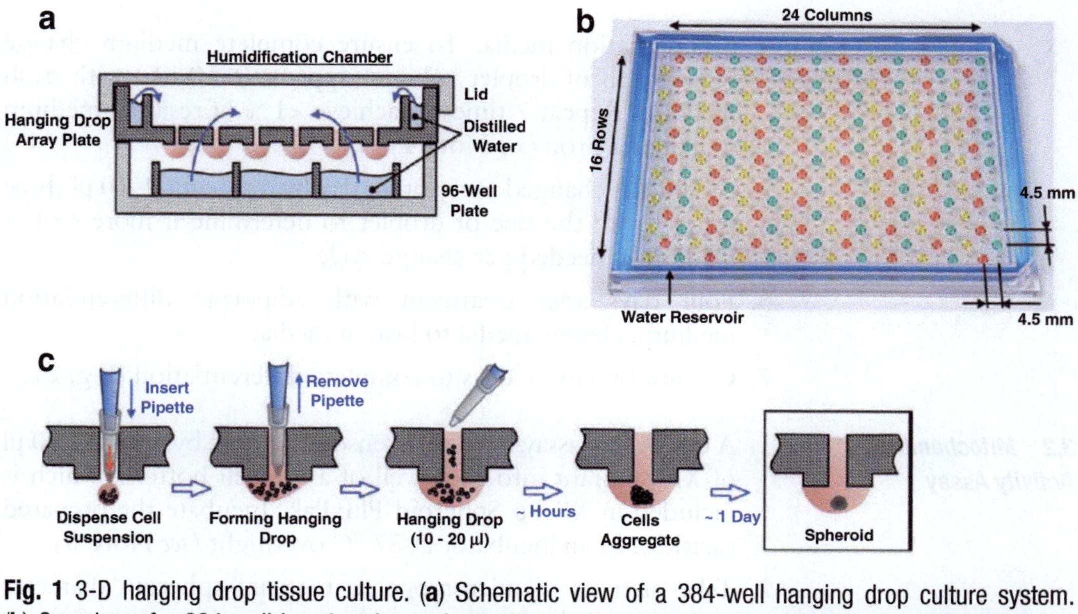

Fig. 1 3-D hanging drop tissue culture. (**a**) Schematic view of a 384-well hanging drop culture system. (**b**) Overview of a 384-well hanging drop culture plate. (**c**) The procedure for spheroid formation. Reprinted from Tung et al. [5]

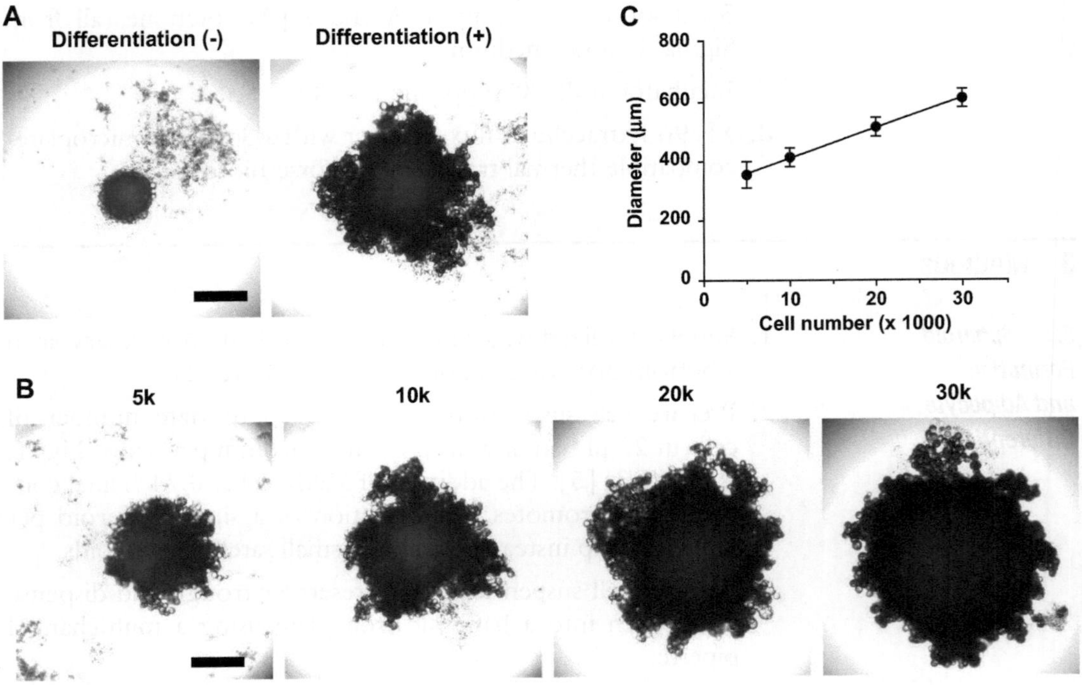

Fig. 2 Cell number- and differentiation-dependent regulation of spheroid size. (**a**) A spheroid of preadipocytes (20 K cells) without adipogenic mix (*left*), adiposphere with adipogenesis. Scale = 200 μm. (**b**) Increased size of adipospheres with increased number of preadipocytes used. Scale = 200 μm. (**c**) Linear relationship between the diameter of adipospheres and the initial number of preadipocytes used. Mean ± SEM. $N = 6$

differentiation media. To ensure complete medium change replace half of droplet volume (approx. 9–10 uL) with fresh medium. Repeat 7 times to achieve <1 % of residual medium in hanging drop (*see* **Note 4**).

5. Medium is changed every other day by replacing 9–10 μl three times. Note the size of droplet to determine if more or less medium is needed per change cycle.
6. Four days after treatment with adipocyte differentiation medium, change media to insulin media.
7. Culture two more days to complete differentiation (Fig. 2).

3.2 Mitochondrial Activity Assay

1. A day before assay, prepare a sensor cartridge by adding 200 μl of XF calibrant into each well of a 96-well bottom, which is included in XFe96 Spheroid FluxPak. Incubate the prepared cartridge in an incubator at 37 °C overnight (*see* **Note 5**).
2. Take pictures of microscope to measure spheroid diameter under inverted microscope.

3. Add mitochondrial assay media into wells of an Ultra-Low attachment multiwell plate. Place the plate below the drop plate with hanging spheroids. Let a hanging spheroid drop down by the addition of 30 μl assay medium (*see* **Notes 6** and 7).
4. Wash spheroids a few times with mitochondria assay media (*see* **Note 8**).
5. Add 165 μl Assay medium into a XFe96 spheroid microplate in a XFe96 spheroid FluxPak.
6. Cut a tip and aspirate 10 μl with a spheroid. Transfer spheroids into the XFe96 spheroid microplate (*see* **Note 9**). Do not transfer spheroids into wells at corners, instead, add only Assay medium.
7. Incubate the XFe96 spheroid microplate in the incubator at 37 °C for 1 h.
8. Dispense 25 μl of oligomycin, FCCP, and antimycin A/rotenone solution into injection ports A, B, and C in the sensor cartridge (*see* **Note 10**).
9. Launch WAVE program in XFe96 Extracellular flux analyzer system. Fill out plate design according to the transferred spheroids into the XFe96 spheroid microplate. Well volume is set to 175 μl.
10. Set measurement cycles after 12 min of equilibration. For all of the cycles, set 3 min for mixing and measurement. There are 5 cycles before injection as basal activity level. After injection of oligomycin from port A, run 12 cycles. Then inject FCCP from port B followed by 5 cycles. After injection of antimycin A and rotenone, 7 cycles last to end of the measurement (Fig. 3).

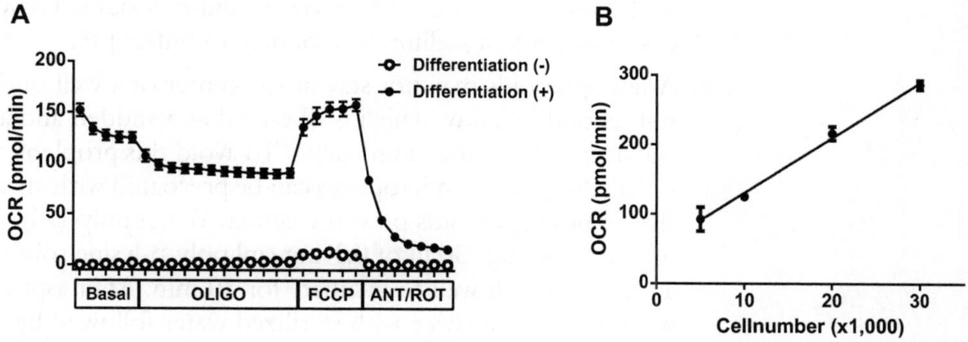

Fig. 3 Differentiation- and cell number-dependent oxygen consumption rate (OCR). (**a**) Differentiated adipocytes (*closed circles*) demonstrate a robust OCR compared to preadipocytes (*open circles*). With the use of a proton ionophore, FCCP, maximal OCRs under uncoupled states are observed. Oligomycin, an inhibitor of complex V, ATP synthase blocks ATP-synthesizing mitochondrial respiration. The use of both Antimycin A, a complex III inhibitor and rotenone, a complex I inhibitor, completely shuts down mitochondrial respiration. (**b**) Adiposphere size (cell number)-dependent increase of basal mitochondrial respiration. Mean ± SEM

4 Notes

1. Substrates and their optimal concentrations to be used in mitochondrial assays may vary depending on the cell type and the metabolic pathways of researcher's interest.
2. To avoid water flowing out by tilting, hot 1 % agarose is poured to solidify after cooling or wet Kim. Or wet Kimwipes will be placed.
3. Appropriate volume for single hanging drop culture is 20–30 μl. Spheroids tend to adhere to the side of the plate when the culture volume is less than 15 μl and fall down to the bottom when more than 35 μl.
4. Medium change can be automated by using a liquid handler CyBi-Well together with an Adapter Standard 384.SQW6 and CyBi-TipTray 96 (all from CyBio, Jena, Germany).
5. The sensor cartridge will be ready to use after 4-h incubation, but overnight incubation is recommended to stable data.
6. As an alternative way, spheroids are collected at once to spin down by using Perfecta3D Spheroid Transfer Tool from 3D Biomatrix. Centrifugation speed may vary depending on cell type and spheroid size. It is also possible to pick up a spheroid from the plate by manually aspirating with a cut tip.
7. If standard cell culture plates are used to collect spheroids, they may adhere to the bottom of the plates quickly. If spheroids are collected into a microtube, they may adhere each other.
8. It is important to reduce bicarbonate ion contained in standard DMEM by washing with mitochondrial assay medium because pH is measured during assay to calculate Extracellular acidification (ECAR). XF Assay Medium Modified DMEM does not contain sodium bicarbonate to buffer pH.
9. A few spheroids may not stay in the center of a well during a mitochondrial assay, which is observed as a sudden and artificial decrease of a measured value. To avoid this problem, wells of XFe96 spheroid microplates can be precoated with an adhesive to hold spheroids onto the center. When poly-D-lysine is used for coating, 30 μl of 0.1 mg/ml poly-D-lysine solution is added into each well to incubate for 20 min. After aspiration, wells are washed twice with sterilized water followed by complete drying up.
10. The concentrations of these antibiotics are optimized for specific types of cells.

Acknowledgment

S.T. has stock options in 3D Biomatrix, a company commercializing the hanging drop plate technology. This research was funded by NIH R01DK095137 (T.H.C.).

References

1. Chun TH, Hotary KB, Sabeh F et al (2006) A pericellular collagenase directs the 3-dimensional development of white adipose tissue. Cell 125(3):577–591
2. Khan T, Muise ES, Iyengar P et al (2009) Metabolic dysregulation and adipose tissue fibrosis: role of collagen VI. Mol Cell Biol 29(6):1575–1591
3. Diculescu I, Stoica M (1970) Fluorescence histochemical investigation on the adrenergic innervation of the white adipose tissue in the rat. J Neurovisc Relat 32(1):25–36
4. Moraes C, Labuz JM, Leung BM et al (2013) On being the right size: scaling effects in designing a human-on-a-chip. Integr Biol 5(9):1149–1161
5. Tung YC, Hsiao AY, Allen SG et al (2011) High-throughput 3D spheroid culture and drug testing using a 384 hanging drop array. Analyst 136(3):473–478

Chapter 18

Culture and Sampling of Primary Adipose Tissue in Practical Microfluidic Systems

Jessica C. Brooks, Robert L. Judd, and Christopher J. Easley

Abstract

Microfluidic culture of primary adipose tissue allows for reduced sample and reagent volumes as well as constant media perfusion of the cells. By continuously flowing media over the tissue, microfluidic sampling systems can more accurately mimic vascular flow in vivo. Quantitative measurements can be performed on or off chip to provide time-resolved secretion data, furthering insight into the dynamics of the function of adipose tissue. Buoyancy resulting from the large lipid storage capacity in this tissue presents a unique challenge for culture, and it is important to account for this buoyancy during microdevice design. Herein, we describe approaches for microfluidic device fabrication that utilize 3D-printed interface templating to help counteract cell buoyancy. We apply such methods to the culture of both isolated, dispersed primary adipocytes and epididymal adipose explants. To facilitate more widespread adoption of the methodology, the devices presented here are designed for user-friendly operation. Only handheld syringes are needed to control flow, and devices are inexpensive and disposable.

Key words Adipose tissue, Microfluidics, 3D printing, Cell culture, Explants, Secretion, Hormones, Glycerol, Insulin signaling, Free fatty acids

1 Introduction

Microfluidic systems have been shown to offer exclusive advantages over traditional methods for chemical and biochemical analysis [1, 2]. Systems can be engineered to integrate multiple steps of standard sample preparation and analysis onto one microdevice, and some devices even furnish analytical tools that are unavailable in their macro-scale counterparts [3, 4]. Recent work in cellular co-culture and organ-on-a-chip platforms have confirmed that the scale of microfluidics is ideal to culture and interrogate tissues [5, 6], and some have even reconstituted human biological function on microdevices that outperform animal models [7]. Others have shown that microfluidics is useful for studying pancreatic islets [8–10], and our group has contributed by developing easy-to-use microdevices for islet stimulation and secretion sampling [11–13].

Jun Wu (ed.), *Thermogenic Fat: Methods and Protocols*, Methods in Molecular Biology, vol. 1566, DOI 10.1007/978-1-4939-6820-6_18, © Springer Science+Business Media LLC 2017

Despite the inherent advantages available in microfluidic cell culture, fewer studies [14] have focused on miniaturizing adipose tissue culture and sampling. Recently, we have demonstrated that by customizing the cell-to-chip interfaces to counteract buoyancy, both isolated adipocytes [15] and adipose tissue explants [13] can be adapted to miniaturized systems. These devices have permitted adiponectin secretion quantification and even time-resolved glycerol secretion measurements on small amounts of endocrine tissue. With these miniaturized methods, a single mouse can provide enough tissue for over an order of magnitude more measurements when compared to macro-scale culture methods, validating a role for microfluidics in adipose tissue biology.

In this chapter, we present detailed approaches for adipose tissue culture and sampling on microfluidic devices. Chip fabrication approaches utilize 3D-printed interface templating to help counteract cell buoyancy for both dispersed primary adipocytes and epididymal adipose explants. To enhance simplicity of use, devices are designed for user-friendly operation with handheld syringes, thus requiring no electrical interfacing or fluidic valving. These inexpensive and disposable devices should be more adaptable for routine use by non-experts, helping to extend the benefits of microscale analysis to more laboratories that study adipocyte biology.

2 Materials

2.1 Photolithography for Silicon Wafer Master Fabrication

1. 4″ Silicon wafers, N-type, <100>, 500 +/− 25 μm, 1SP (Polishing Corporation of America).
2. Transparency design (FineLine Imaging).
3. Glass plate for design mounting.
4. Oven (*see* **Note 1**).
5. Rotary shaker.
6. Glass dish for cleaning 4″ wafers.
7. Wafer tweezers (150 SA WAFER, Vomm Germany).
8. 1.0 M sulfuric acid.
9. Distilled, deionized H_2O (ddH_2O).
10. Hot plate.
11. N_2 gas line or supply.
12. Spin coater (WS-400BZ-6NPP/Lite, Laurell Technologies).
13. Spin coater alignment tool (Thingiverse.com, EasleyLab) (*see* **Note 2**).
14. SU-8 negative photoresist and developer (MicroChem).

15. Photolithography system with ultraviolet light source for photocuring of SU-8 (*see* **Note 3**).
16. UV blocking safety glasses.
17. Isopropyl alcohol (IPA).

2.2 PDMS Microchip Devices

1. Sylgard 184® elastomer base and curing agent (Dow Corning).
2. Silicon wafers.
3. Tygon tubing (0.02 inch ID, 0.06 inch OD).
4. Blunt-ended needles (22 G).
5. 60-mL plastic, luer lock syringes.
6. 3-mL luer lock syringes.
7. Plasma cleaner (Harrick; basic cleaner, PDC-32G).
8. Vacuum desiccator and pump.
9. Aluminum foil.
10. Double-sided tape.
11. Oven or heated rocker (*see* **Note 1**).
12. 100-mm Petri dishes.

2.3 Manually Fabricated Interface Templates

1. Polymer casting resin (Smooth-On, Inc.; Smooth-Cast 310).
2. Silicone oil.
3. Disposable biopsy punches with plunger (Militex®, prod # 15110).
4. Vacuum desiccator and pump.
5. Plasma cleaner (Harrick; basic cleaner, PDC-32G).

2.4 3D-Printed Interface Templates

1. 3D sketching software (e.g., SketchUp, OpenSCAD) or pre-designed files (*see* **Note 2**).
2. 3D printer (MakerBot Replicator 2).
3. Printer filament (Makerbot, polylactic acid).

2.5 Fat Pad Isolation and Culture

1. Disposable culture tubes, 12 × 75 mm.
2. 3 mL luer lock, sterile syringe.
3. Polypropylene mesh, 210 μm (Spectra Mesh, prod # 145880).
4. Sterile 96-well plate with lid.
5. 18G × 1 1/2 inch needles.
6. 2 mm disposable biopsy punch with plunger.
7. Distilled, deionized H_2O (ddH_2O).
8. Phosphate-HEPES Buffer: bovine serum albumin (BSA), d-glucose, HEPES, NaCl, $CaCl_2$–2 H_2O, $MgSO_4$–7 H_2O, KH_2PO_4, Na_2HPO_4, pH: 7.2–7.4.

9. Adipocyte Media: nystatin suspension (10,000 units/mL), penicillin-streptomycin (10,000 U/mL), 100× minimal essential media, nonessential amino acids solution (MEM NEAA), fetal bovine serum (FBS) heat inactivated, dulbecco's modified eagle medium (DMEM, low glucose), dulbecco's modified eagle medium (DMEM, without glucose and phenol red), fatty acid free bovine serum albumin (FAF BSA), sodium pyruvate, L-glutamine, nalgene® rapid-flow™ filter units, and bottle top filters (PES membrane, 0.2 μm pore size, sterile).
10. Serum Media: Add 6 mL of FBS, nystatin, penicillin-streptomycin, and MEM NEAA to 476 mL of low glucose DMEM. Filter into sterile container and store at 4 °C.
11. Serum Free Media: Add 6 mL of nystatin, penicillin-streptomycin, and MEM NEAA; 5 mL of sodium pyruvate; 10 mL of L-glutamine; and 1 g of FAF BSA to 467 mL of glucose and phenol red free DMEM.
12. Collagenase solution: Weigh 10 mg of collagenase type I into a 15 mL conical tube and add 8 mL of phosphate-HEPES buffer. Mix gently by inversion and place in water bath.

3 Methods

3.1 Photolithography for Silicon Wafer Master Fabrication (See Note 4)

1. Carefully remove new wafer and place in glass dish. Add in enough 1.0 M sulfuric acid to just cover the wafer in the dish (Fig. 1a).
2. Secure the dish on a rotary shaker. Turn on shaking at ~250 rpm for 30 min.
3. At the end of the acid wash, remove the wafer from the dish with wafer tweezers and rinse both the wafer and the dish with ample ddH_2O.
4. Return the wafer to the dish and cover it with ddH_2O. Turn on shaking at ~250 rpm for 30 min.
5. Remove the wafer from the dish with wafer tweezers, carefully dry with N_2 gas, and place in an oven at 200 °C for 15 min (Fig. 1b).
6. Remove the wafer from the oven, place on a clean Kimwipe, and allow it to cool to room temperature.
7. Center wafer on spin coater with a wafer alignment tool (*see* **Note 2**). Pour SU-8 slowly onto the wafer's center to avoid air bubbles (~1 mL SU-8 per inch of substrate diameter). Choose the grade of SU-8 (based on viscosity) which corresponds to desired channel height.
8. Spin the wafer at 500 rpm with an acceleration of 100 rpm s^{-1} for 30 s (Fig. 1c).

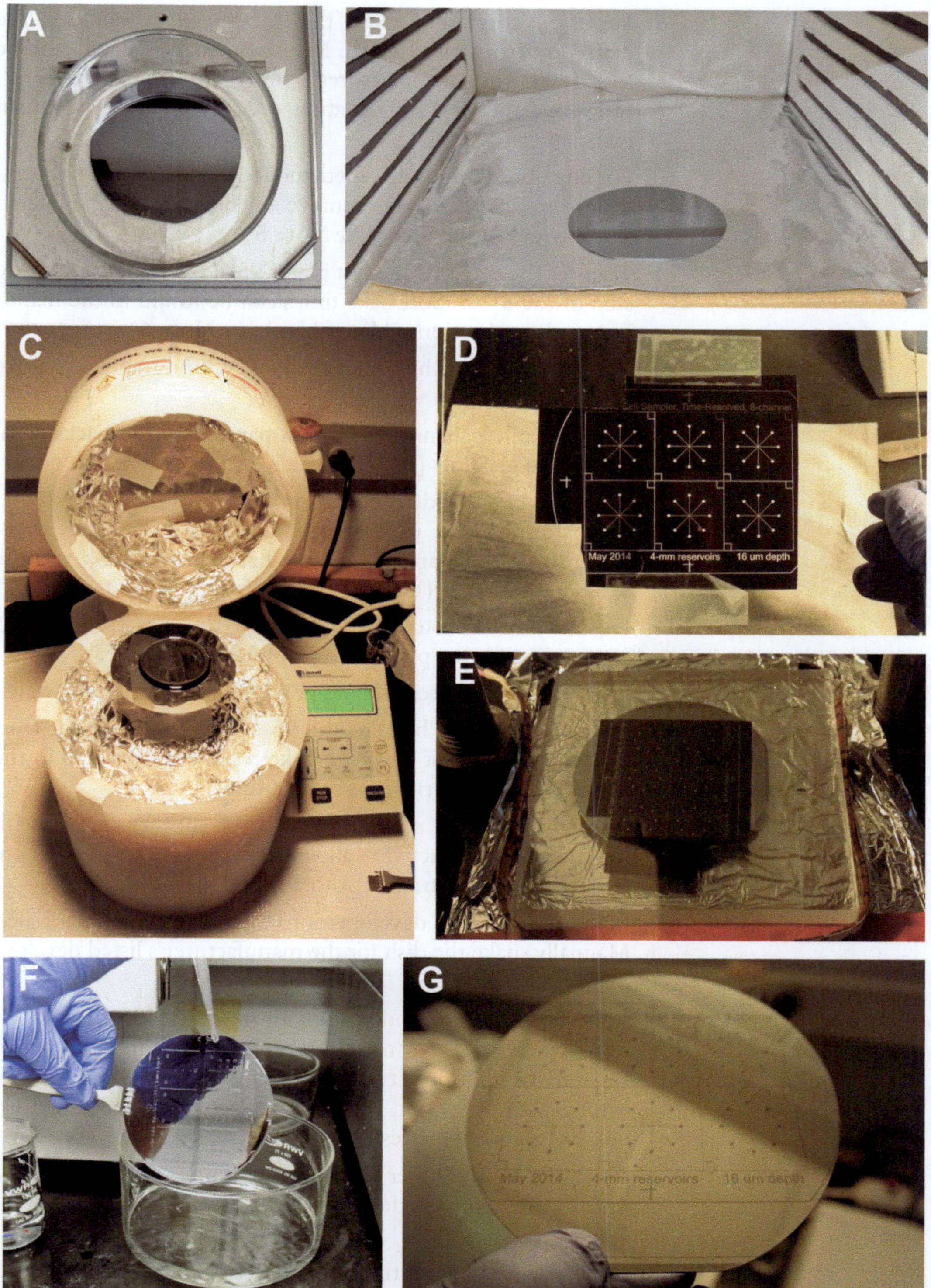

Fig. 1 Photolithography steps. (**a**) Wafer is acid cleaned on a rotary shaker. (**b**) Wafer is baked at 200 °C to remove residual water. (**c**) Spin coating of SU-8 onto clean wafer. (**d**) Microchannel mask transparency is taped onto a dust-free glass slide. (**e**) Wafer and design placed into UV lithography system for exposure step. (**f**) Photoresist development step. (**g**) Finilazed SU-8 master wafer, ready for PDMS soft lithography

9. For the second spin cycle, the spin speed depends on the desired microchannel depth. Exact spin coating details can be found in the manufacturer's instructions (Microchem). However, the first spin coating cycle (**step 8**) will remain constant.
10. For edge bead removal, a methanol-soaked Kimwipe or cleanroom wipe can be used to remove any bulk SU-8 accumulated at the edges of the wafer (*see* **Note 5**).
11. Transfer the wafer to the hot plate (95 °C) for the soft baking step of the SU-8. Soft bake times can be found in the manufacturer's instructions (Microchem) (*see* **Note 6**).
12. Place the wafer in the center of the UV lithography exposure platform.
13. Tape the microchannel mask transparency to the backside of the square glass tile (Fig. 1d).
14. Gently place the slide centered on the top of the wafer so that the patterned side of the mask gently contacts the SU-8. Ensure that there are no air gaps between the mask and the SU-8 on the wafer (Fig. 1e).
15. Expose the wafer to UV light using the exposure settings appropriate for the photolithography system in use.
16. Lift the glass slide from the wafer without also lifting the wafer.
17. Transfer the wafer back to the 95 °C hot plate for the post exposure bake. Post exposure bake times can be found in the manufacturer's instructions (*see* **Note 6**).
18. Place the wafer on a clean Kimwipe or another surface with low thermal conductivity, allow it to passively cool to room temperature, then transfer it to a round glass dish.
19. Pour enough SU-8 developer into the dish to cover the wafer. Manually shake the dish for the manufacturer's listed development time.
20. Rinse both the dish and wafer with fresh developer for ~10 s, followed by an IPA rinse to remove any residual developer. Intricate or high-density channel designs may require forceful rinsing with developer, which can be accomplished using disposable transfer pipettes (Fig. 1f).
21. Carefully dry the wafer with N_2 gas. If any white film forms, rinse again with fresh developer for ~10 s followed by another IPA rinse.
22. Once dry and clean, the wafer is ready for use as a PDMS channel master (Fig. 1g).

3.2 PDMS Microchip Devices

1. Mix precursors of polydimethylsiloxane (PDMS) base and curing agent (10:1 respectively) in a plastic weigh boat. For a single silicon wafer, ~45.0 g total of PDMS is sufficient.

Stir PDMS precursors with a disposable wooden spatula until sufficiently mixed.

2. Place weigh boat into a degassing chamber and allow to degas until the polymer is bubble free. While the PDMS is degassing, rinse the silicon wafer with methanol and dry with N_2 gas.
3. Cut a square of aluminum foil large enough to form a boat that contains the silicon wafer. Place a strip of double-sided tape on the bottom of the wafer and attach it to the center of the aluminum foil boat. Gently push the wafer flat onto the center of the foil square to ensure attachment, and carefully fold up the foil around the edges of the wafer to form a boat-like structure that can hold liquid PDMS precursors.
4. Slowly pour the degassed PDMS precursor liquid onto the top of the wafer. If any additional bubbles are formed in the polymer, a second degassing can be applied to remove them.
5. At this point, interface templates can be placed over the appropriate structures in the PDMS to define the reservoir regions above channel designs (*see* Figs. 2b, 3b, and 4). Degas one final time to remove any air trapped between the wafer and the template.

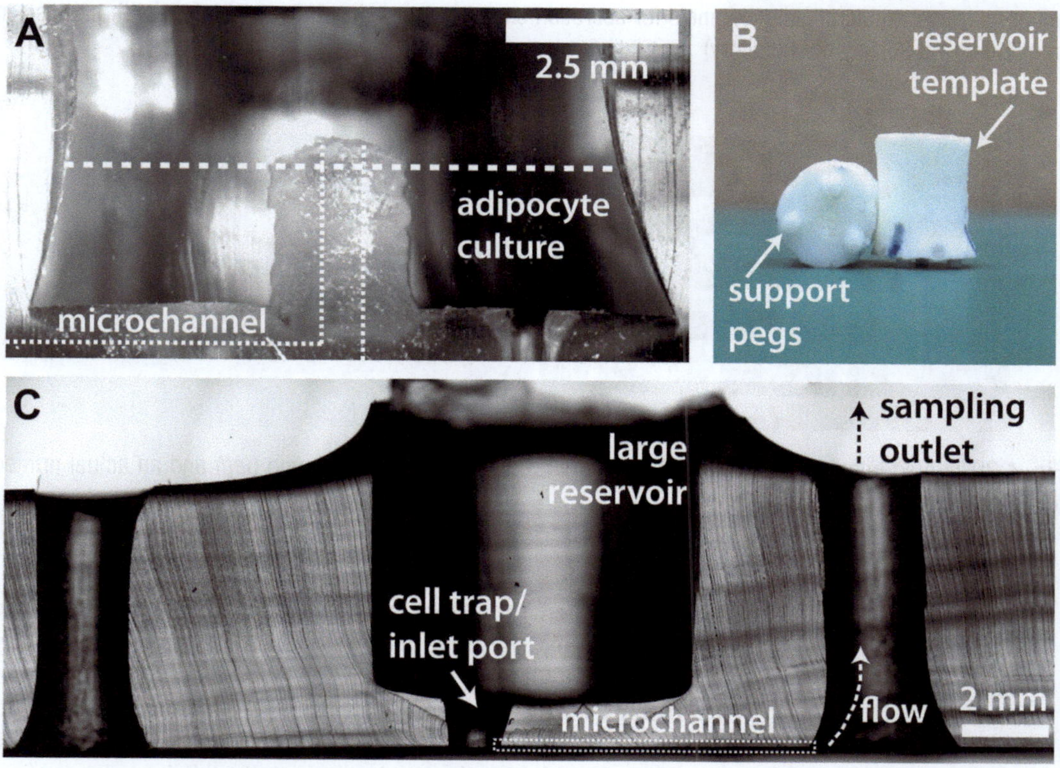

Fig. 2 Cross-sections of typical devices. (**a**) A cross-section of a device made with a manually fabricated interface template. (**b**) Examples of manually fabricated interface templates, showing support pegs that define reservoir height. (**c**) A cross-section of a device made with a 3D-printed interface template

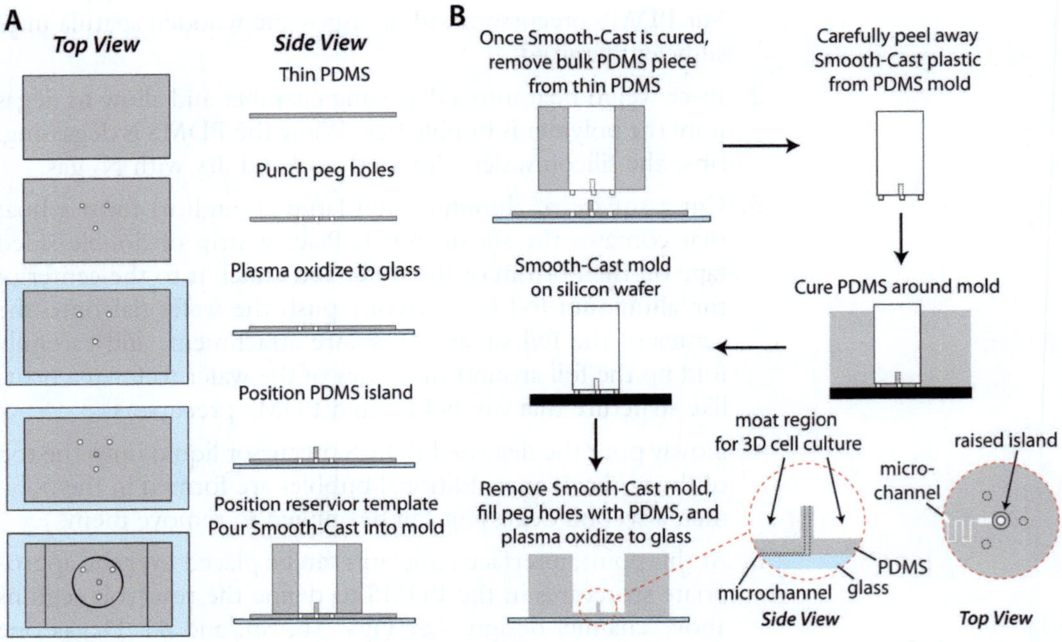

Fig. 3 Manual fabrication method for the Smooth-Cast 310 interface template. (**a**) The PDMS mold for the insert serves as an initial proxy for the final device, including the raised island and moat region customized for culture of primary adipocytes in collagen. (**b**) The resultant template is used to define a customized macro-to-micro interface to microfluidic channels for secretion sampling (reproduced from Ref. [15] with permission from The Royal Society of Chemistry)

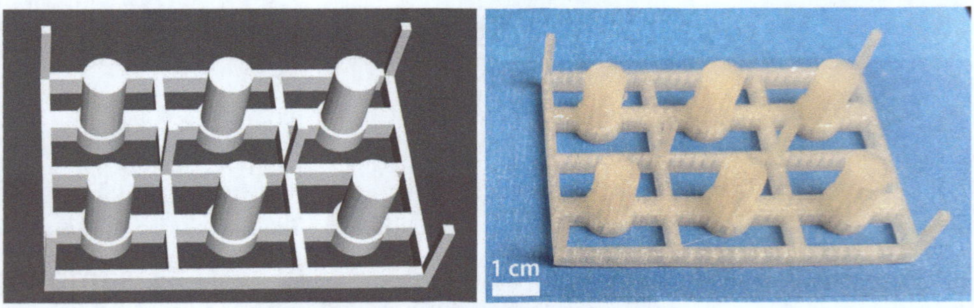

Fig. 4 3D-printed interface templates. Shown here are a 3D software rendering (*left*) and an actual printed template (*right*) for explant culture

6. Place the wafer, templates, and PDMS in an oven to cure the polymer. Adjust the final placement of the template by gently pushing the structures into the PDMS, and avoid lifting them from the PDMS to prevent air bubbles. Curing times around 2 h are sufficient for ovens at or above 50 °C. PDMS can be cured at 60–80 °C when using manually fabricated interface templates, but the oven should be kept near 50 °C when using 3D-printed interface templates (*see* **Note 7**).

7. Carefully peel and remove the aluminum foil from the PDMS and the wafer. A razor blade can be used to remove excess pieces of foil or PDMS from the wafer's edges.
8. Gently begin to lift the PDMS from the wafer, starting at the edges and working around the circumference of the wafer. Once the edges have been released, slowly lift the bulk of the PDMS from the wafer, minimizing the pressure applied to the wafer, which can be easily broken.
9. Rinse the wafer with methanol and dry with N_2 gas. Return the water to a 100 mm petri dish for storage until the next usage.
10. Cut the bulk PDMS down to the single unit device pieces with a razor, making sure to leave clean edges for optimum plasma bonding to the glass floor.
11. With the channel sides facing up, begin punching inlet and outlet holes in the microchip. Rinse thoroughly with methanol and dry with N_2 gas. Apply and remove Scotch tape to the channeled side of the chip several times to remove any remaining dust particles or debris. If there are multiple chips to be cleaned and bonded, tape can be left on the cleaned chips to prevent any contamination.
12. Glass slides for plasma bonding should also be rinsed with methanol, dried with N_2 gas, and taped clean before use. The glass slides serve as the channel floors and help to stabilize the device.
13. Remove any remaining tape and place both the glass slide and the PDMS chip into the plasma oxidation chamber. Surfaces that will be bonded should be uncovered and facing up (i.e., channeled side up).
14. Plasma oxidize the slide and chip for 1 min at medium power. Carefully remove the slide and chip, making sure not to touch the activated surfaces. Grasp the chip by the sides and turn so that the channels are facing the glass slide. In a rolling motion, apply the PDMS to the middle section of the glass slide. Rolling, instead of simply placing the whole piece of PDMS down at once, helps prevent trapping of air bubbles between the chip and the glass slide.
15. Ensure that there are no air bubbles by viewing the chip contact through the glass slide. Medium pressure by hand can be applied to ensure full contact. If pressed too firmly, channels could collapse and become bonded to the glass as well. Do not disassemble the glass and PDMS to retry the bonding, as covalent bonds between the two surfaces will have already formed, and the chip will be damaged beyond use. Regions of thin PDMS membranes where interface templates have been removed often need to be gently pushed with a blunt probe to achieve full contact for bonding to the glass substrate. Cross-sections of typical devices are shown in Fig. 2.

16. Once the chip is bonded, fill the main reservoir with serum free media containing BSA. Let the chips incubate with the solution for at least 1 h. The BSA will adsorb to the freshly cleaned surfaces and channels to prevent nonspecific binding during experiments. This step should be carried out the day of plating collagen-based adipocytes or the day of experiments for tissue explants. Do not allow the chips to dry out once they have been treated with media.
17. Before plating adipocytes and collagen on the chip, remove residual serum free media from the reservoir. Immediately begin plating the cells in collagen as per Subheading 3.5.
18. If adipose tissue explants are to be used on the chips, remove excess media from the reservoir and transfer the explants via forceps to the punched inlet. Before adding back fresh media, place the 3D-printed explant trap into the reservoir to prevent the explant from floating. Continue to Subheading 3.7 for microfluidic sampling procedures.

3.3 Manually Fabricated Interface Templates (Fig. 3)

1. Punch 1 mm diameter holes in a 1 mm thick slab of PDMS to create support legs for the template mold. The thickness of the slab will define the depth of PDMS between the large reservoir and microfluidic channels in the resulting microfluidic device, and this parameter can be altered to accommodate experimental needs (Fig. 3a).
2. Plasma oxidize the slab and a glass microscope slide. Place the PDMS in the center of the slide to covalently bond the two (Fig. 3a).
3. Punch a 2–3 mm diameter plug out of PDMS and place the plug in the center of the mold form. This plug will define the raised island in the large reservoir (Fig. 3a).
4. The body of the mold should be punched into a thicker piece of PDMS (~2 cm height). This piece should be centered over the support legs to reversibly seal (not plasma oxidize) the two pieces of PDMS and complete the mold (Fig. 3a, bottom).
5. Measure pre-polymer solutions of Smooth-Cast (part A and part B), and degas the unmixed components for 10 min. Once degassed, combine 100 μL of each part (1:1 by volume) and pipette the solution into the small support peg section of the PDMS mold; degas the prepolymer once more. Due to the small size of the support pegs, it is imperative that the Smooth-Cast is properly degassed to prevent a brittle mold foundation. After degassing, mix the remaining Smooth-Cast (~1 mL), and carefully add it to the bulk of the mold. Degas to remove any additional air trapped in the body of the mold, and allow to cure overnight.

6. To remove the Smooth-Cast template, lift the thick slab of PDMS from the thin slab on the glass slide. Carefully peel the template from the PDMS mold, being careful to not tear the PDMS or break the support pegs. Any extra pieces of PDMS stuck to the template can be removed with an X-Acto™ knife or a blunt-ended needle (Fig. 3b, top).
7. The manually fabricated interface template is now ready to template adipocyte culturing reservoirs (Fig. 3b) into PDMS devices during curing of PDMS on the SU-8 patterned wafer (*see* Subheading 3.2, **step 5**). Examples and usage of these templates were reported by Godwin et al. [15], and an image of two templates is shown in Fig. 2b.

3.4 3D-Printed Interface Templates

1. To use predesigned interface templates, choose the appropriate 3D part from those available. Refer to our laboratory's profile on Thingiverse website (*see* **Note 2**).
2. Alternatively, templates can be customized or modified, exploiting the rapid prototyping feature of 3D printing systems. Using appropriate 3D CAD software (e.g., SketchUp, OpenSCAD), design 3D structure of the custom interface template.
3. Print the interface template using a 3D printer. Templates and parts shown in this chapter were printed using a MakerBot Replicator 2 with PLA filament. 3D renderings and sample prints of these templates are shown in Fig. 4.
4. The 3D-printed interface template is now ready to template adipocyte culturing reservoirs into PDMS devices during curing of PDMS on the SU-8 patterned wafer (*see* Subheading 3.2**, step 5**).

3.5 Adipocyte Digestion and Microfluidic Culture

1. Immediately before surgery, make the collagenase solution. Place the collagenase, phosphate-HEPES buffer, and serum media in 37 °C water bath.
2. Remove epididymal fat pads from C57BL/6J male mice and transfer to 4 mL of pre-warmed phosphate-HEPES buffer.
3. Place the extracted fat into a 60 mm dish to record the wet weight of the tissue.
4. Transfer the extracted fat into a 2 mL snap cap tube. Mince the tissue in the tube for 2 min using surgical scissors.
5. For every 1 g of fat, add 2 mL of collagenase to the sample. Pipette the suspension up and down several times, then place the tube in a rocking water bath set at 120 rpm and 37 °C.
6. Allow the tissue to digest for 30 min in the water bath. To ensure adequate digestion, pipette the suspension up and down every 10 min during the digestion.

7. Prepare a funnel with 210 μm Spectra mesh. Pipette the digested tissue over the filter and into a 5 mL glass test tube. Wash the cells with 3–4 mL of phosphate-HEPES buffer.
8. Centrifuge the cell suspension at 900 rpm (123 × *g*) for 6 min at 4 °C.
9. Remove the infranatant with an 18G 1 1/2 inch needle and small syringe.
10. Wash cells with an additional 3–4 mL of phosphate-HEPES buffer and repeat centrifuging and infranatant removal.
11. After the second wash, add 3–4 mL of serum media and centrifuge at 1000 rpm (150 × *g*) for 6 min at room temperature.
12. Remove infranatant and add 3–4 mL of serum media.
13. Place the tube with freshly washed cells in the CO_2 incubator for 30 min. Lightly cover the top of the tube with Parafilm.
14. Immediately before the incubation is complete, mix 4.5 mL of collagen, 450 μL 10× MEM, and 3–5 μL of 1.0 M NaOH. Upon addition of the 10× MEM, the solution will turn yellow. Due to the viscosity of collagen, make sure to vortex thoroughly. When adding the NaOH, a light pink end color is desired. The freshly mixed collagen will begin to harden and can only be used for ~30 minutes before it becomes too viscous for plating.
15. Pipette the fat cells floating on the top layer of the serum media and transfer to a 2 mL snap cap tube. Record the total volume of cells collected.
16. Add the collagen directly to the cell solution (10 μL of cells to 45 μL collagen). Pipette the solution to ensure homogeneity and begin plating (55 μL/microwell). Gently shake the chips to cover the surface evenly with cells. Continuously pipette to mix the stock suspension during the plating process.
17. After all the microdevice wells have been plated with cell suspension and collagen, place the chips into the CO_2 incubator for 60 min. The collagen and cell mixture will have solidified in this time, and serum media can be slowly added to each well. An example of digested and reconstituted adipose tissue cultured on a microdevice is shown in Fig. 5a.
18. Allow the cells to sit in the incubator overnight, at which time the cells and devices will be ready for microfluidic secretion sampling.

3.6 Explant Culture on Microdevices

1. Place phosphate-HEPES and serum media in a water bath at 37 °C before the surgery.
2. Remove epididymal fat pads from C57BL/6J male mice and transfer to 4 mL of pre-warmed phosphate-HEPES buffer.
3. Transfer the fat pads to a 60 mm dish containing fresh phosphate-HEPES buffer. Carefully cut away any large veins from the tissue and discard them.

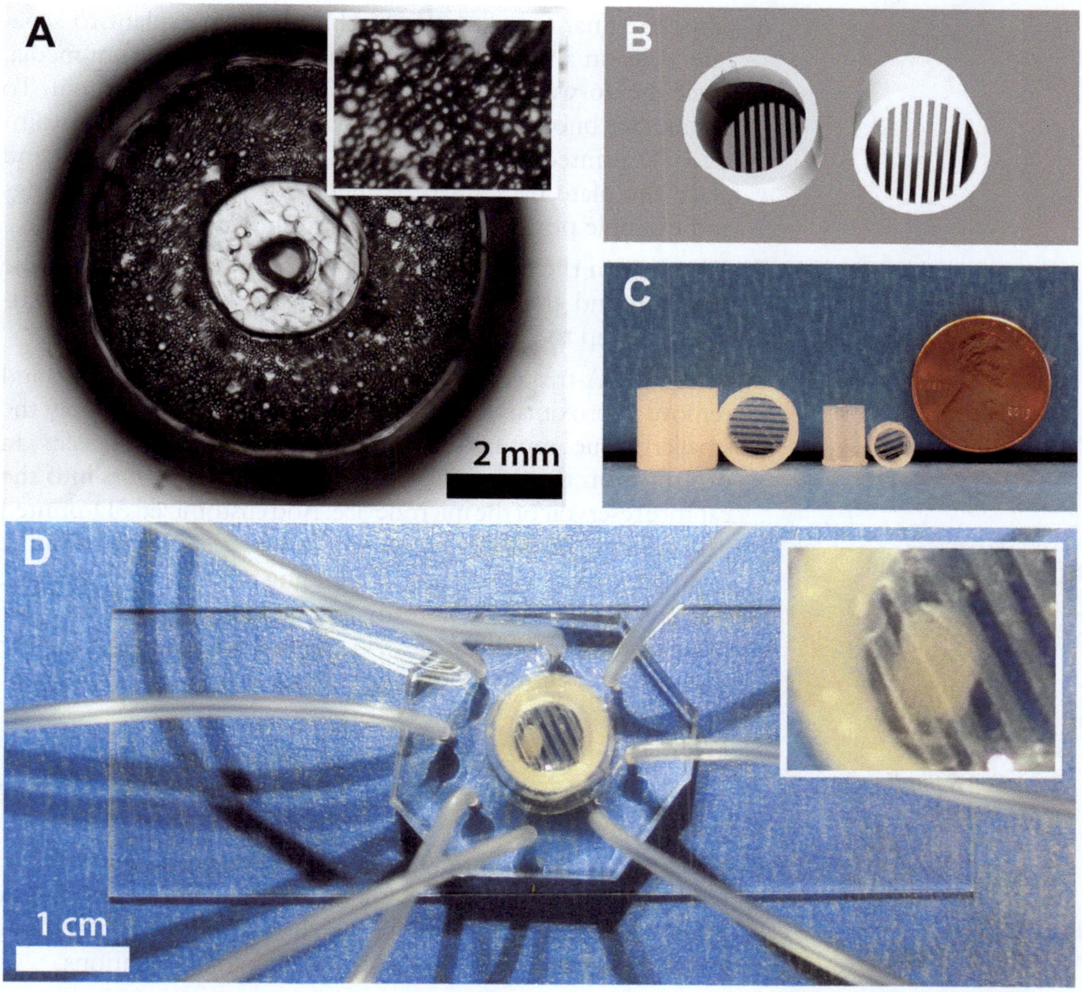

Fig. 5 Adipocytes cultured on 8-channel microfluidic sampling devices. (**a**) *Top-down view* image of digested and reconstituted adipose tissue cultured on a microdevice made with a 3D-printed interface template. In*set* shows zoomed version of adipocytes in collagen. (**b**) 3D renderings and (**c**) example prints of adipose tissue explant traps. (**d**) Image of a trapped adipose tissue explant cultured on a microdevice and held in place with an explant trap. *Inset* shows zoomed version of trapped explant

4. Punch 2 mm diameter plugs of tissue using biopsy punches with plungers. Transfer each explant as it is punched into a glass test tube with phosphate-HEPES buffer. Continue punching until sufficient tissue has been collected or the remaining tissue is insufficient for accurate punching.
5. Centrifuge the explants at 1000 rpm (150×*g*) for 3 min. Remove the infranatant using a 18-gauge needle and a 3 mL syringe. Replace with 4 mL of phosphate-HEPES buffer. Repeat this process two additional times, replacing the buffer with 4 mL of serum media.

6. After the final rinse, transfer each individual explant to separate wells on a 96-well plate containing 200 μL serum media. Place the 96-well plate into the CO_2 incubator for 30 min. To counteract buoyancy during preparative culture, add customized 3D-printed explant traps (*see* **Note 2**) into each well and place the plate back into the incubator [13]. 3D renderings and example prints of these traps are shown in Fig. 5b.
7. To maintain the explants, wash with fresh serum media in the morning and the evening. Explants can be used for experiments for up to 7–10 days if maintained in this way.
8. Select a BSA-treated device (*see* Subheading 3.2, **step 16**), and remove approximately 90 % of the BSA buffer so that only the smaller tissue capture reservoir contains buffer. Transfer an explant from the 96-well plate using surgical tweezers into the capture reservoir of the microdevice. Add customized 3D-printed explant trap (*see* **Note 2**) to counteract buoyancy, then add the appropriate starting solution, depending on the experiment. At this point, devices will be ready for microfluidic secretion sampling of the trapped explant. An example of a trapped adipose tissue explant cultured on a microdevice is shown in Fig. 5c.

3.7 Adipocyte Secretion Sampling in Microfluidic Devices

1. Pre-warm all solutions and media in a water bath at 37 °C.
2. Wash cells three times with fresh serum media, and allow them to incubate in the media for at least 1 h.
3. While the cells are bathing in fresh media, prepare treatment media buffer. If the final sample solution will be used in a homogeneous assay, serum free media without phenol red can be used to prevent readout issues with colored solutions.
4. A variety of treatments and timings can be applied to these small portions of tissue on the microdevice, depending upon the experiment of choice. As an example, we have shown in Fig. 6 that glycerol secretion is strongly dependent upon insulin and glucose concentrations at physiologically relevant levels. Importantly, up to an order of magnitude more experiments can be carried out on single mouse using this microfluidic system, compared to standard sampling techniques.
5. When the cultured adipocytes or explants are ready for sampling, insert Tygon tubing into the outlet holes on the chip. On the opposite end of the tubing, slide in a blunt-ended needle at least 3–4 mm.
6. Attach a 60 mL syringe to the luer lock end of the blunt needle, with the syringe already set to the starting volume. Carefully apply vacuum to the chip via the syringe and place a lock or stopper on the plunger of the syringe to prevent its movement.
7. After sampling has completed, remove the syringe from the tubing to stop the vacuum.

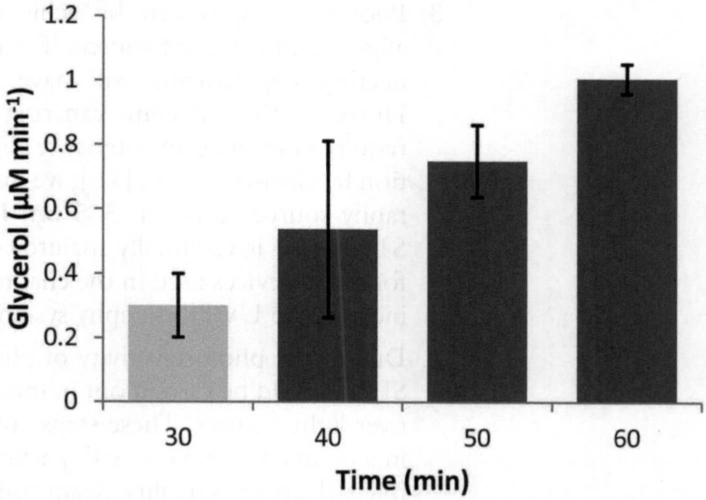

Fig. 6 Time-resolved glycerol secretion from an adipose tissue explant using 3D-templated reservoirs and a 3D-printed tissue trap on a microfluidic device under continuous, vacuum-driven flow. After switching from high insulin and glucose (2 nM insulin, 19 mM glucose) to low insulin and glucose (50 pM insulin, 3.5 mM glucose) solutions at the 30 min time point, glycerol secretion rates increased. Average data is shown for three explants, with error bars representing standard deviations

8. To collect the sample from the tubing, place a 3 mL luer lock syringe on the end of the tubing, but do not apply any vacuum. Have an opened, 1.7 mL snap cap tube ready for the transfer. Gently pull the tubing from the chip with the 3 mL syringe still in place. Hold the open end of the tubing in the snap cap tube and slowly apply pressure from the 3 mL syringe to the tubing to transfer the sample volume.
9. Depending on the assay format, samples can be immediately analyzed or placed in a freezer at −20 °C until future analysis.

4 Notes

1. Convection ovens will provide more consistent results with photolithography or device curing, but a standard laboratory oven should suffice for these devices.
2. 3D-printed templates and accessory devices mentioned in this chapter are currently available for free download from our laboratory's profile on the Thingiverse website (http://www.thingiverse.com/; EasleyLab). Parts are available in STL format and are print-ready. Specifically, we printed these parts using a Makerbot Replicator 2, a desktop 3D printer that retails for less than $2000.

3. Photolithography can be achieved using a standard mask aligner and UV light source, if available. Most electrical engineering departments will have access to such a system. However, these systems can cost greater than $20,000 and require continued maintenance. Following the 2014 publication by Groisman et al. [16], we built a customized UV lithography source based on 365 nm LEDs, which cost less than $1000 and is essentially maintenance free. All SU-8 masters for microdevices used in the chapter were fabricated using this inexpensive UV lithography system.
4. Due to the photosensitivity of photoresist, all steps involving SU-8 should be carried out in minimal light or with UV filters over light fixtures. These steps are also preferably carried out in a clean environment with particulate filtered air, although a microelectronics quality cleanroom is not required.
5. Once the wafer is coated with SU-8, be mindful to not contact the wafer with the tweezers or any other objects in areas where channel designs will be patterned.
6. To avoid thermally compromising the SU-8/silicon adhesion and reduce bubble formation, it may be necessary to start the hot plate at a lower temperature then ramp slowly to the required baking temperature (e.g., 95 °C). Cooling ramps will also help maintain the adhesion, thus rapid cooling should be avoided.
7. PDMS precursors can be cured at higher temperatures than 50 °C and are often cured at 80 °C in our laboratory and others. However, when using 3D-printed interface templates, the oven temperature must be kept below the glass transition temperature of the PLA material, which is around 60–65 °C. This precaution will avoid warping of the 3D-printed template structure.

Acknowledgments

Support for the work was provided by the National Institutes of Health (R01 DK093810) as well as by the Department of Chemistry and Biochemistry and the College of Science and Mathematics at Auburn University. The authors would like to thank Mark D. Holtan and Tesfagebriel Hagos for assistance with device fabrication photographs. We also extend thanks to Dr. Leah A. Godwin for initiating the work in this area.

References

1. Lee JN, Park C, Whitesides GM (2003) Solvent compatibility of poly(dimethylsiloxane)-based microfluidic devices. Anal Chem 75:6544–6554
2. Whitesides GM, Ostuni E, Takayama S et al (2001) Soft lithography in biology and biochemistry. Annu Rev Biomed Eng 3:335–373
3. Easley CJ, Karlinsey JM, Landers JP (2006) On-chip pressure injection for integration of infrared-mediated DNA amplification with electrophoretic separation. Lab Chip 6:601–610
4. Kim J, Johnson M, Hill P, Gale BK (2009) Microfluidic sample preparation: cell lysis and

nucleic acid purification. Integr Biol 1: 574–586

5. Gao Y, Majumdar D, Jovanovic B et al (2011) A versatile valve-enabled microfluidic cell co-culture platform and demonstration of its applications to neurobiology and cancer biology. Biomed Microdevices 13:539–548
6. Halldorsson S, Lucumi E, Gómez-Sjöberg R, Fleming RMT (2015) Advantages and challenges of microfluidic cell culture in polydimethylsiloxane devices. Biosens Bioelectron 63:218–231
7. Bhatia SN, Ingber DE (2014) Microfluidic organs-on-chips. Nat Biotechnol 32:760–772
8. Dishinger JF, Reid KR, Kennedy RT (2009) Quantitative monitoring of insulin secretion from single islets of Langerhans in parallel on a microfluidic chip. Anal Chem 81:3119–3127
9. Adewola AF, Wang Y, Harvat T et al (2010) A multi-parametric islet perifusion system within a microfluidic perifusion device. J Vis Exp 35:e1649
10. Yi L, Wang X, Dhumpa R et al (2015) Integrated perfusion and separation systems for entrainment of insulin secretion from islets of Langerhans. Lab Chip 15:823–832
11. Easley CJ, Rocheleau JV, Head WS, Piston DW (2009) Quantitative measurement of zinc secretion from pancreatic islets with high temporal resolution using droplet-based microfluidics. Anal Chem 81:9086–9095
12. Godwin LA, Pilkerton ME, Deal KS et al (2011) Passively operated microfluidic device for stimulation and secretion sampling of single pancreatic islets. Anal Chem 83:7166–7172
13. Brooks JC, Ford KI, Holder DH et al (2016) Macro-to-micro interfacing to microfluidic channels using 3D-printed templates: Application to time-resolved secretion sampling of endocrine tissue. Analyst 141:5714–5721
14. Moraes C, Labuz JM, Leung BM et al (2013) On being the right size: scaling effects in designing a human-on-a-chip. Integr Biol 5: 1149–1161
15. Godwin LA, Brooks JC, Hoepfner LD et al (2015) A microfluidic interface for the culture and sampling of adiponectin from primary adipocytes. Analyst 140:1019–1025
16. Erickstad M, Gutierrez E, Groisman A (2015) A low-cost low-maintenance ultraviolet lithography light source based on light-emitting diodes. Lab Chip 15:57–61. doi:10.1039/c4lc00472h

Chapter 19

Using Thermogenic Beige Cells to Identify Biologically Active Small Molecules and Peptides

Ling Wu and Bin Xu

Abstract

Incorporating molecular libraries in chemical biology screenings in cultured cells has been successfully used for gene discovery in many cellular processes. It has the unique potential to uncover novel mechanisms of complex cellular biology through the screening of small molecules and protein biologics in relevant cell-based assays. Recent development in the understanding and generation of thermogenic adipocytes provides opportunities for potential anti-obesity therapeutics discovery. In this chapter, we describe screening methods using thermogenic beige cells to identify novel compounds and peptides that activate adipocyte thermogenesis.

Key words Thermogenic adipocyte, Beige cell, Small molecule and peptide screening, Quantitative PCR

1 Introduction

In light of the worldwide diabetes and obesity pandemic, new pharmacological strategies are needed to balance energy homeostasis. The activities of brown and beige fat cells reduce metabolic dysfunction—including diabetes and obesity—in mice and correlate with leanness in humans. Brown adipocytes have been located in specific areas and express constitutively high levels of thermogenic genes, whereas inducible "brown-like" adipocytes, also referred to as beige adipocytes, develop in white fat depots in response to various activators [1]. The significant advances in our understanding of beige adipocytes made in the last few years provide excellent opportunities for small molecule lead compound and peptide hormone discoveries [2–8].

Thermogenic beige cells can respond to a variety of stimuli, including cold temperatures, sympathetic stimulation, small molecules drugs such as thiazolidinediones (TZDs), and peptide hormones, such as the exercise-induced peptide irisin. The adaptive thermogenesis activities in adipocytes can be measured by increased mRNA levels of browning marker genes. The most prominent

Jun Wu (ed.), *Thermogenic Fat: Methods and Protocols*, Methods in Molecular Biology, vol. 1566,
DOI 10.1007/978-1-4939-6820-6_19,

members of browning genes include *Ucp-1*, *Cidea*, *Pgc-1α* [2, 9]. Additional members of browning genes that have been frequently cited include *Prdm16*, *Elovl3*, and *aP2* [9, 10]. The levels of these validated browning genes can be used as readouts for developing assays in screening small molecules and peptides as activators of adipocyte thermogenesis. Increased thermogenesis in adipose tissue may also be reflected in increased protein expression levels of these marker genes [11].

A number of peptide hormones or secreted protein factors have been reported to regulate energy expenditure in adipose tissue (for recent reviews, *see* [3, 5]). These protein factors (such as exercise-induced myokine irisin, heart or brain-related natriuretic peptides, bone morphogenic protein (BMP) family proteins, fibroblast growth factor Fgf21, and neuropeptide orexin) come from diverse sources. For a number of these factors, their functions remain to be elucidated. Agents promoting adipose browning are not limited to peptide hormones. Recent studies showed that several small molecules strongly induced the browning of white fat. The small molecules reported have structures as diverse as alkaloid berberine [11], amino acid β-aminoisobutyric acid [12], catecholamines [13], inorganic nitrate [14], and thyloid hormone T4 [15]. With increased interest and research activities on adipose biology, more activators of adaptive thermogenesis, small molecules, or secreted peptide hormones are likely to be discovered.

Taking advantage of the rapid developments in the field of thermogenic adipose biology, we used clonal thermogenic beige cell lines to identify novel activators. In this chapter, we will describe a set of molecular cell biology assays to identify small molecules and peptides as browning agents. These methods may be readily expanded for high-throughput assays. Major assays and procedures are illustrated in a schematic flow chart in Fig. 1.

2 Materials

2.1 Preparation of Small Molecules and Peptides

1. Cell culture grade DMSO.
2. Amicon Ultra-15 Centrifugal Filter Units or equivalent.
3. Ultrafree-MC Centrifugal Filter Devices with 0.22 μM pore size or equivalent.

2.2 Cell Culture and Cell Treatment

1. Mouse beige adipose cell line X9 (Generously provided by Prof. Bruce Spiegelman, Harvard Medical School and Dana-Farber Cancer Institute; [2]).
2. 24-well cell culture plate.
3. Dulbecco's phosphate-buffered saline (DPBS).
4. 0.25 % Trypsin–EDTA.

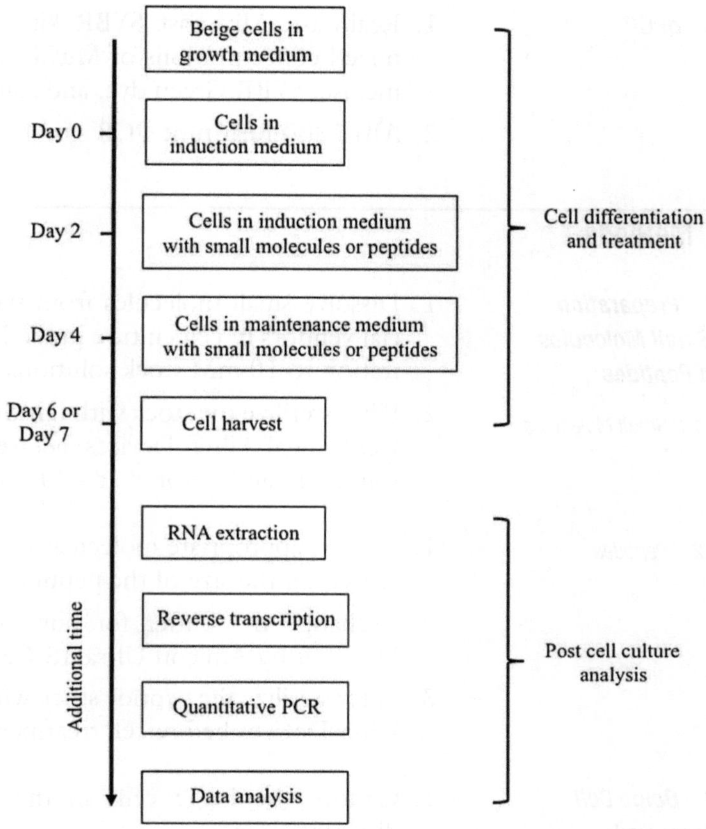

Fig. 1 Schematic experimental flow chart of cell differentiation, treatment, and posttreatment cell analysis. Optimal time and duration for beige cell differentiation and drug treatment are suggested. Compositions of various media are specified in Subheading 2

5. Cell growth medium: Prepare DMEM/F-12 GlutaMAX (Thermo Fisher Scientific) or equivalent with 15 % fetal bovine serum (FBS) and 1 % penicillin/streptomycin.
6. Cell induction medium: Prepare DMEM/F-12 GlutaMAX with 10 % FBS, 1 % penicillin/streptomycin, 5 μM dexamethasone, 0.5 μg/mL insulin, 0.5 mM isobutylmethylxanthine (IBMX), 1 μM rosiglitazone, and 1 nM triiodothyronine (T3).
7. Cell maintenance medium: Prepare DMEM/F-12 GlutaMAX with 10 % FBS, 1 % penecillin/streptomycin, 0.5 μg/mL insulin, and 1 nM T3.

2.3 mRNA Purification and cDNA Synthesis

1. TRIzol, chloroform, isopropanol, 100 % ethanol, diethylpyrocarbonate (DEPC).
2. SuperScript III First-Strand Synthesis System (Thermo Fisher Scientific) or equivalent.
3. Water bath or programmable thermal cycler.
4. UV-Vis spectrophotometer.

2.4 qPCR

1. RealMasterMix Fast SYBR kit or equivalent contains optimized concentrations of $MgCl_2$, dNTPs, SpeedAB Taq polymerase, SYBR Green dye, and stabilizers.
2. ABI 7300 Real-time PCR system or equivalent.

3 Methods

3.1 Preparation of Small Molecules and Peptides

3.1.1 Small Molecules

1. Dissolve small molecules from synthetic sources or commercial vendors in cell culture grade DMSO to bring the concentration to 10 mM stock solutions.
2. Filter sterilize the stock with DMSO-compatible Ultrafree-MC Centrifugal Filter Devices before cell treatment. The stock solutions can be stored at −20 °C.

3.1.2 Peptides

1. Choose appropriate molecular weight cutoff of the filter units (based on the size of the peptides).
2. Exchange the buffer for purified recombinant peptides to DPBS using Amicon Ultra-15 Centrifugal Filter Units.
3. Filter sterilize the peptide stock with Ultrafree-MC Centrifugal Filter Devices before cell treatment.

3.2 Beige Cell Culture and Differentiation

1. Culture the beige cells in the cell growth medium until differentiation.
2. Plate cells in 24-well plates at the density of 1.5×10^5 cells/well (*see* **Note 1**).
3. Next day (counted as Day 0 in cell differentiation; *see* Fig. 1 for the timeline of cell culture and differentiation events): To initiate differentiation, remove cell growth medium and add 500 μL induction medium to the confluent beige cells.
4. Replace with fresh induction medium on Day 2. Start targeted small molecules or peptides treatment.
5. After 4 days' cell induction, beige cells were switched to cell maintenance medium containing the same small molecule or peptide treatment.
6. Harvest treated beige cells and control cells on Day 6 or Day 7 (*see* **Note 2**).

3.3 Beige Cell Treatment

1. On day 2, cells were switched from growth medium to 500 μL differentiation medium for induction.
2. Small molecules or peptides were added on Day 2 in the cell differentiation medium. Starting concentrations of 1–10 μM for small molecules, or 20 nM–1 μM for peptides are recommended (*see* Fig. 2 for examples). Gently mix and incubate the cells with the treatment for 48 h.

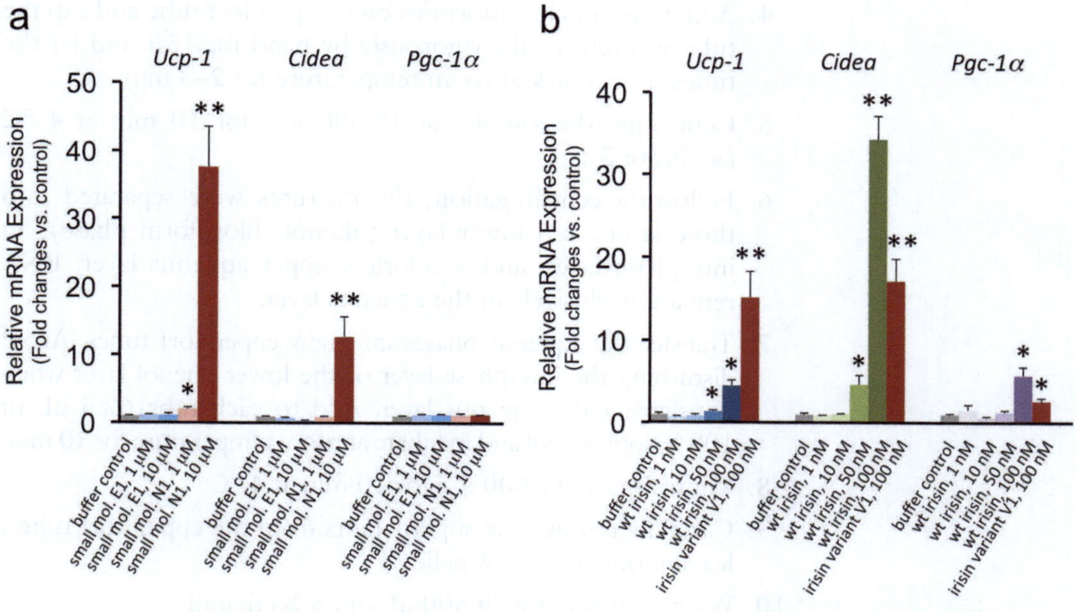

Fig. 2 Illustration of browning gene induction in beige X9 cells with the treatment of small molecules (**a**) or peptides (**b**). Total RNA was isolated from the differentiated beige cells with or without treatment and mRNA levels of brown-fat-like genes (*Ucp-1*, *Cidea*, and *Pgc-1α*) were quantified by qPCR. Values are mean ± s.e.m. ($n = 3$ replicates). Significant differences compared with vehicle controls (bars shown in *gray color*) are indicated by *, $p < 0.05$ and **, $p < 0.01$ (assessed by Student's t-test). (**a**) Small molecule E1 serves as a negative control. Small molecule N1 serves as a positive lead compound identified. (**b**) Wild-type recombinant human irisin serves as a positive control as a peptide hormone. Irisin variant V1 serves as an example of peptide activator identified. *Ucp-1* and *Cidea* typically show the strongest induction among a list of browning gene markers. Induction of *Pgc-1α* often occurs with the treatment of browning activators, but not always [6, 12]

3. On day 4, switch cell differentiation medium to maintenance medium with the same treatment.
4. On day 6 or day 7, remove the culture medium for total RNA extraction or storing the 24-well plates at −80 °C for follow-up processing.

3.4 Total RNA Extraction

The following procedures were modified and adapted based on initial protocol suggested by the vendor.

1. Remove the growth medium from the culture plates. Rinse the cells once with ice-cold DPBS. Add 125 μL TRIzol reagent directly to each well in the culture plates to lyse the cells.
2. Homogenize the cell lysates by pipetting the lysates up and down several times.
3. Transfer the lysates to clean, autoclaved eppendorf tubes. Vortex thoroughly and then incubate the homogenized samples at room temperature for 5 min to permit complete dissociation of the nucleoprotein complexes.

4. Add 25 μL of chloroform to each eppendorf tube and cap the tube securely. Shake vigorously by hand for 15 s and let the tubes sit in a rack at room temperature for 2–3 min.
5. Centrifuge the samples at 12,000 × *g* for 10 min at 4 °C (*see* **Note 3**).
6. Following centrifugation, the mixtures were separated into three layers: red lower layer (phenol-chloroform phase), an interphase layer, and a colorless upper aqueous layer. RNA remains exclusively in the aqueous layer.
7. Transfer the aqueous phases into new eppendorf tubes. Avoid disturbing the interphase layer or the lower phenol layer when transferring the aqueous layer. Add to each tube 62.5 μL of 100 % isopropanol and incubate at room temperature for 10 min.
8. Centrifuge at 12,000 × *g* for 10 min at 4 °C.
9. Carefully remove the supernatants from the eppendorf tubes, leaving only the RNA pellets.
10. Wash each pellet with 500 μL of 75 % ethanol.
11. Vortex the sample briefly, then centrifuge the tube at 7500 × *g* for 5 min at 4 °C. Aspirate the supernatants.
12. Air-dry the RNA pellets for 5–10 min (*see* **Note 4**).
13. Resuspend the RNA pellets in RNase-free water (DEPC-treated water).
14. Incubate the eppendorf tubes in a water bath or a programmable thermal cycler at 55–60 °C for 10–15 min to increase RNA yields.
15. Estimate RNA concentrations by UV-Vis spectrophotometer (*see* **Notes 5** and **6**). Proceed to downstream applications or store the extracted RNA samples at −80 °C.

3.5 cDNA Synthesis

cDNA synthesis is performed with the SuperScript III First-Strand Synthesis System for RT-PCR. The following procedures were modified and adapted based on initial protocol suggested by the vendor.

1. Place 0.2 mL PCR micro-tubes on ice. For each tube, add 0.5 μg total RNA (*see* **Note 7**), 1 μL 50 μM oligo$(dT)_{20}$ primer (*see* **Note 8**), 1 μL 10 mM dNTPs, and RNase-free water up to 10 μL.
2. After sealing each reaction, vortex the mixtures gently. Centrifuge briefly to collect components at the bottom of the reaction tubes.
3. Incubate the tubes at 65 °C for 5 min, then quickly cool down on ice for 2 min.
4. Spin down briefly and add the following reagents in each tube: 2 μL 0.1 M DTT, 1 μL (40 units) RNaseOUT (*see* **Note 9**), 1 μL

(200 units) of SuperScript III RT, 4 μL 25 mM $MgCl_2$, 2 μL 10 × RT buffer (200 mM Tris–HCl (pH 8.4), 500 mM KCl).

5. Mix gently and collect by brief centrifugation.
6. Incubate at 50 °C for 50 min and inactivate the reaction by heating at 85 °C for 5 min.
7. Chill on ice and spin down briefly. Add 1 μL (2 Units) of *E. coli* RNase H to each tube and incubate the tubes at 37 °C for another 20 min (*see* **Note 10**).
8. The first strand cDNA can be stored at −20 °C until further use as a template for amplification in quantitative PCR.

3.6 Quantitative PCR (qPCR)

qPCR is carried out with the RealMasterMix SYBR ROX system.

1. Dilute cDNA products with 80 μL buffer containing 10 mM Tris–HCl (pH 8.0) and 0.1 mM.
2. Mix gene-specific forward and reverse primer pair, each at the concentration of 2 pmol/μL.
3. Prepare the following mixture in a 96-well plate: 10 μL SYBR Green Mix (2×), 2 μL cDNA, 3 μL primer pair mix, and 5 μL PCR-grade water. Make sure all the reagents are completely thawed and well mixed before use to maintain data reproducibility (*see* **Note 11**).
4. Mix the reactions well and spin if needed. Seal and label the 96-well plate (*see* **Note 12**).
5. Set up the experiment using the following PCR program on ABI 7300 real-time PCR system.
   (a). 50 °C 2 min, 1 cycle.
   (b). 95 °C 10 min, 1 cycle (*see* **Note 13**).
   (c). 95 °C 15 s followed by 60 °C 1 min, 40 cycles.
6. After the PCR program is completed, remove the tubes from the PCR machine. (Optional) The PCR specificity can be examined by 3 % agarose gel using 5 μL from each reaction.
7. Put the tubes back in ABI 7300 system and perform dissociation curve analysis with the saved copy of the setup file.
8. Analyze the real-time PCR results with ABI 7300 system SDS software. Check to see if there is any bimodal dissociation curve or abnormal amplification plot.
9. Calculate relative mRNA expression levels according to the 2(-Delta Delta C(T)) Method [16].

3.7 Interpreting, Validating, and Expanding the Screening Experiments

Like any biological assays, controls are important for validating the screening results. 18S ribosomal RNA is one of the most commonly used internal controls for qPCR assays. Known small molecule and peptide activators such as those described in Subheading 1 may serve as positive controls for this method. DMSO, DPBS buffers, or any non-activators may be used as negative controls. The selection of the threshold and the number of hits to be considered for further analysis may be guided by (a) the statistical significance of the increased folds of the browning gene markers, (b) elevation of multiple browning gene markers, (c) dose-response analysis with varying concentrations of treatment, and (d) secondary screening results, such as Western blot analysis of the marker protein UCP-1 or cell-based immunostaining with antibodies against UCP-1 protein [11, 12]. Current 24-well plate format can be readily adapted and expanded to 96-well plate format.

4 Notes

1. Cell confluency is critical for beige cell differentiation. Low density of beige cells may affect its optimal differentiation efficiency.
2. Observe the beige cell differentiation status daily. Visually check cell morphology changes (such as lipid drops).
3. RNA pellets formed are colorless after centrifugation. The gel-like pellets will be at the side and on the bottom of the eppendorf tubes. To avoid RNA loss, it is recommended that a pipetman be used to aspirate the supernatant instead of using a vacuum apparatus.
4. Do not over-dry the RNA pellets to avoid solubility issues in the follow-up step of RNA resuspension in water.
5. For using TRIzol method to extract RNA, a low A260/280 ratio (<1.6) may suggest lower purity of the extracted RNA. Contaminations with proteins, phenol, and those having strong absorbance at or near 280 nm may affect the measurements. The following steps may help to improve the quality of RNA. First, add sufficient amount of TRIzol in the step of cell lysis. Typically, 0.2 mL of chloroform per 1 mL of TRIzol reagent (or at this ratio) was added during the step of cell lysate homogenization. Secondly, do not disturb the interphase layer or phenol-chloroform layer when transferring the upper aqueous layer. Perform extra wash with 75 % ethanol after isopropanol precipitation of RNA. An effective but relatively expensive way is to use QIAGEN RNase column to clean up RNA after TRIzol extraction.
6. Low yield of RNA and genomic DNA contamination are two potential problems during the RNA preparation. Both incomplete homogenization and RNA degradation can cause low yields of RNA. Sufficient volume of TRIzol and complete homogenization of cells may facilitate the shearing of the genomic DNA and fully releasing of RNA. Reasons for RNA degradation are often difficult to pinpoint. All the material for

the experiment should be RNase-free. Try not to repeat freezing-and-thawing cycles of RNA samples. Adding DNase is the most direct way to remove DNA from RNA preparations.

7. Total RNA amount may be ranged from 1 pg to 5 μg. Optimal RNA quantity has to be determined empirically.
8. Oligo(dT) is recommended over random hexamers or gene-specific primers to enrich poly A+ mRNA in RT-PCR.
9. RNaseOUT recombinant RNase inhibitor prevents target RNA degradation from ribonuclease and other RNases contamination during the RNA preparation.
10. RNase H facilitates removing RNA template from the cDNA:RNA hybrid molecules after first-strand synthesis and therefore it may increase the PCR yields.
11. To minimize errors during pipetting, mix cDNA template, SYBR Green Mix, and water and aliquot into PCR plate with pre-added primers. Perform every reaction in triplicates.
12. Using a sealer scraper may facilitate tight sealing of adhesive film to a plate. Make sure any labeling on the PCR tubes does not interfere the instrument light path for detection. Clean the surface of the transparent sealer with 70 % ethanol.
13. The initial denaturation time of 1–2 min is typically sufficient. For DNA templates with a high GC content, denaturation time may be increased up to 15 min.

Acknowledgment

We thank Professor Bruce Spiegelman (Harvard Medical School and Dana-Farber Cancer Institute) for generously providing beige cell X9 line. We thank Prof. Jun Wu (University of Michigan Life Sciences Institute) for stimulating discussions. This work is in part supported by Virginia Tech New Faculty Start-up Funds, the National Institute of Food and Agriculture of the USDA (Program No. VA-135992), and grants from Diabetes Action Research and Education Foundation (DAREF), Commonwealth Health Research Board (CHRB) and Virginia Tech Center for Drug Discovery (VTCDD).

References

1. Rosen ED, Spiegelman BM (2014) What we talk about when we talk about fat. Cell 156(1–2):20–44. doi:10.1016/j.cell.2013.12.012
2. Wu J, Boström P, Sparks LM, Ye L, Choi JH, Giang AH, Khandekar M, Virtanen KA, Nuutila P, Schaart G, Huang K, Tu H, van Marken Lichtenbelt WD, Hoeks J, Enerbäck S, Schrauwen P, Spiegelman BM (2012) Beige adipocytes are a distinct type of thermogenic fat cell in mouse and human. Cell 150(2):366–376. doi:10.1016/j.cell.2012.05.016
3. Harms M, Seale P (2013) Brown and beige fat: development, function and therapeutic potential. Nat Med 19(10):1252–1263. doi:10.1038/nm.3361

4. Spiegelman BM (2013) Banting lecture 2012: regulation of adipogenesis: toward new therapeutics for metabolic disease. Diabetes 62(6):1774–1782. doi:10.2337/db12-1665
5. Wu J, Cohen P, Spiegelman BM (2013) Adaptive thermogenesis in adipocytes: is beige the new brown? Genes Dev 27(3):234–250. doi:10.1101/gad.211649.112
6. Rao RR, Long JZ, White JP, Svensson KJ, Lou J, Lokurkar I, Jedrychowski MP, Ruas JL, Wrann CD, Lo JC, Camera DM, Lachey J, Gygi S, Seehra J, Hawley JA, Spiegelman BM (2014) Meteorin-like is a hormone that regulates immune-adipose interactions to increase beige fat thermogenesis. Cell 157(6):1279–1291. doi:10.1016/j.cell.2014.03.065
7. Moisan A, Lee YK, Zhang JD, Hudak CS, Meyer CA, Prummer M, Zoffmann S, Truong HH, Ebeling M, Kiialainen A, Gérard R, Xia F, Schinzel RT, Amrein KE, Cowan CA (2015) White-to-brown metabolic conversion of human adipocytes by JAK inhibition. Nat Cell Biol 17(1):57–67. doi:10.1038/ncb3075
8. Pfeifer A, Hoffmann LS (2015) Brown, beige, and white: the new color code of fat and its pharmacological implications. Annu Rev Pharmacol Toxicol 55:207–227. doi:10.1146/annurev-pharmtox-010814-124346
9. Boström P, Wu J, Jedrychowski MP, Korde A, Ye L, Lo JC, Rasbach KA, Boström EA, Choi JH, Long JZ, Kajimura S, Zingaretti MC, Vind BF, Tu H, Cinti S, Højlund K, Gygi SP, Spiegelman BM (2012) A PGC1-α-dependent myokine that drives brown-fat-like development of white fat and thermogenesis. Nature 481(7382):463–468. doi:10.1038/nature10777
10. Huh JY, Dincer F, Mesfum E, Mantzoros CS (2014) Irisin stimulates muscle growth-related genes and regulates adipocyte differentiation and metabolism in humans. Int J Obes (Lond) 38(12):1538–1544. doi:10.1038/ijo.2014.42
11. Zhang Z, Zhang H, Li B, Meng X, Wang J, Zhang Y, Yao S, Ma Q, Jin L, Yang J, Wang W, Ning G (2014) Berberine activates thermogenesis in white and brown adipose tissue. Nat Commun 5:5493. doi:10.1038/ncomms6493
12. Roberts LD, Boström P, O'Sullivan JF, Schinzel RT, Lewis GD, Dejam A, Lee YK, Palma MJ, Calhoun S, Georgiadi A, Chen MH, Ramachandran VS, Larson MG, Bouchard C, Rankinen T, Souza AL, Clish CB, Wang TJ, Estall JL, Soukas AA, Cowan CA, Spiegelman BM, Gerszten RE (2014) β-Aminoisobutyric acid induces browning of white fat and hepatic β-oxidation and is inversely correlated with cardiometabolic risk factors. Cell Metab 19(1):96–108. doi:10.1016/j.cmet.2013.12.003
13. Nguyen KD, Qiu Y, Cui X, Goh YP, Mwangi J, David T, Mukundan L, Brombacher F, Locksley RM, Chawla A (2011) Alternatively activated macrophages produce catecholamines to sustain adaptive thermogenesis. Nature 480(7375):104–108. doi:10.1038/nature10653
14. Roberts LD, Ashmore T, Kotwica AO, Murfitt SA, Fernandez BO, Feelisch M, Murray AJ, Griffin JL (2015) Inorganic nitrate promotes the browning of white adipose tissue through the nitrate-nitrite-nitric oxide pathway. Diabetes 64(2):471–484. doi:10.2337/db14-0496
15. López M, Varela L, Vázquez MJ, Rodríguez-Cuenca S, González CR, Velagapudi VR, Morgan DA, Schoenmakers E, Agassandian K, Lage R, Martínez de Morentin PB, Tovar S, Nogueiras R, Carling D, Lelliott C, Gallego R, Oresic M, Chatterjee K, Saha AK, Rahmouni K, Diéguez C, Vidal-Puig A (2010) Hypothalamic AMPK and fatty acid metabolism mediate thyroid regulation of energy balance. Nat Med 16(9):1001–1008. doi:10.1038/nm.2207
16. Livak KJ, Schmittgen TD (2001) Analysis of relative gene expression data using real-time quantitative PCR and the 2(-Delta Delta C(T)). Methods 25(4):402–408

INDEX

Jun Wu (ed.), *Thermogenic Fat: Methods and Protocols*, Methods in Molecular Biology, vol. 1566,
DOI 10.1007/978-1-4939-6820-6, © Springer Science+Business Media LLC 2017

MIX
Papier aus verantwortungsvollen Quellen
Paper from responsible sources
FSC® C105338

If you have any concerns about our products,
you can contact us on
ProductSafety@springernature.com

In case Publisher is established outside the EU,
the EU authorized representative is:
Springer Nature Customer Service Center GmbH
Europaplatz 3, 69115 Heidelberg, Germany

Printed by Libri Plureos GmbH
in Hamburg, Germany